Special Publication No. 85

Plasma Source Mass Spectrometry

The Proceedings of the Third Surrey Conference on Plasma Source Mass Spectrometry

University of Surrey, 16th–19th July 1989

Edited by
K.E. Jarvis, A.L. Gray, and J.G. Williams
Royal Holloway and Bedford New College, London

I. Jarvis
Kingston Polytechnic

British Library Cataloguing in Publication Data
Surrey Conference on Plasma Source Mass Spectrometry (*3rd: 1990: Guildford, England*)
Plasma source mass spectrometry.
1. Mass spectrometry
I. Title II. Jarvis, K. E. III. Royal Society of Chemistry
IV. Series
543.0873

ISBN 0-85186-567-4

Published by The Royal Society of Chemistry,
Thomas Graham House, The Science Park, Cambridge CB4 4WF

Printed in Great Britain by Henry Ling Ltd, at the Dorset Press, Dorchester.

INTRODUCTION

The Third Surrey Conference on Plasma Source Mass Spectrometry, held in July 1989, was the latest in a series of highly successful biennial conferences devoted to this exciting new field of analytical science. We are delighted that all of the subject areas covered at the conference are represented here in this proceedings volume. Although still a relatively new technique (reflected in a significant number of fundamental studies), the areas of application of the technique are wide and varied.

The problems of sample introduction are covered in the first two contributions. Firstly the method of laser ablation is considered by Moenke-Blankenburg and co-workers with attention to vapour transport processes. Klaus Dittrich highlights a number of alternative methods of sample introduction with particular reference to the use of electrothermal vaporization.

The reduction of spectral interferences in inductively coupled plasma mass spectrometry (ICP-MS) has received some attention from Diane Beauchemin and Jane Craig with the addition of molecular gasses (added via a sheathing device) into an argon plasma. Although traditionally ICP-MS instruments have used ion detectors in the pulse counting mode, giving very high sensitivity, one instrument company has produced a system whereby analogue detection can be used. Hutton and co-workers discuss the characteristics of this mode of operation. Calibration strategy is an important consideration in any analytical technique. The contribution by Michael Ketterer, John Reschl and Michael Peters addresses the determination of metals in geological samples using a rigorous multivariate calibration scheme.

The application of ICP-MS to the analysis of biological, environmental, geological, metallurgical and nuclear samples is dealt with next. Contributions from the joint winners of the conference poster prize cover the analysis of human serum (Hans Vanhoe and co-workers) and the preparation and analysis of plant materials, by Ed McCurdy. Both papers concentrate on the ultra-trace level determination of a wide variety of elements. A new compound with strongly alkaline properties, has been successfully used for the digestion of mainly plant materials by Toshitsura Cho, Isoko Akabane and Yukio Murakami. The authors demonstrate its applicability with particular reference to the determination of some volatile elements.

A review of Re-Os isotope ratio determinations in geological samples is given by Jean Richardson, Alan Dickin and Robert McNutt. The current instrumental limitations are highlighted and data are presented from a number of applications. The contribution from Jane Moore and co-workers illustrates the application of ICP-MS to trace element analysis of iron materials.

The impact of ICP-MS in the nuclear field is illustrated by two contributions dealing with the determination of U and the actinide elements. In the first by Richard Hall and co-workers, ultra-trace levels of plutonium and uranium have been successfully determined using electrothermal vaporisation as the mode of sample introduction. In the second by Toole *et al.*, several of the actinide elements have been determined in natural reference materials.

In view of the scope of these contributions, we are certain that all those involved in the development and application of inductively coupled plasma mass spectrometry will find these proceedings of great value.

July 1990 Kym Jarvis, Alan Gray, Ian Jarvis and John Williams

CONTENTS

PROCESSES OF LASER ABLATION AND VAPOUR TRANSPORT TO THE ICP

L. Moenke-Blankenburg, M. Gäckle, D. Günther and J. Kammel

Department of Chemistry
The Martin-Luther-University Halle-Wittenberg
Halle 4050
German Democratic Republic

1 INTRODUCTION

Soon after the first report of laser action in ruby in 1960 it was generally recognised that the intense laser output beam, when focused, could be used to vaporise, excite and ionise materials. Over more than 25 years, laser microanalysis[1] based on atomic emission, mass separation, and later on atomic absorption and atomic fluorescence, has been evolving and changing. In the last ten years there have been many new ideas and developments, especially in tandem techniques, such as Laser Micro ICP Atomic Emission Spectroscopy (LM-ICP-AES)[2], and Laser Ablation ICP Mass Spectrometry (LA-ICP-MS)[3], (Figure 1). All modifications of laser micro analysis or laser ablation techniques serve as methods for direct chemical analysis of solids mainly on the micro scale.

The operation of the system is simple. Solid samples of the appropriate size are placed in a glass capped chamber and orientated such that their surface is at the focus of the laser beam (Figure 2). The radiation of a solid state laser with an output energy of about 1 J is focused by the optical system of a microscope (Figure 3) on a selected region of the sample. It evaporates a picogram to microgram amount of substance or ablates microgram to milligram amounts of analytical material. The Ar carrier gas flow is passed over the sample surface and then through a tube to the ICP torch (Figure 2). The gas transports the majority of sample vapour and small amounts of microscopic liquid droplets, solidified or amorphous particles of sizes up to 2μm.

Laser micro analysis has three principal fields of application: (a) it serves as a microchemical method; (b) a method for local analysis, and (c) as a method for distribution analysis. Microchemical analysis of non-conducting, as well as conducting, powders using laser radiation for vaporisation requires very small amounts of substance,

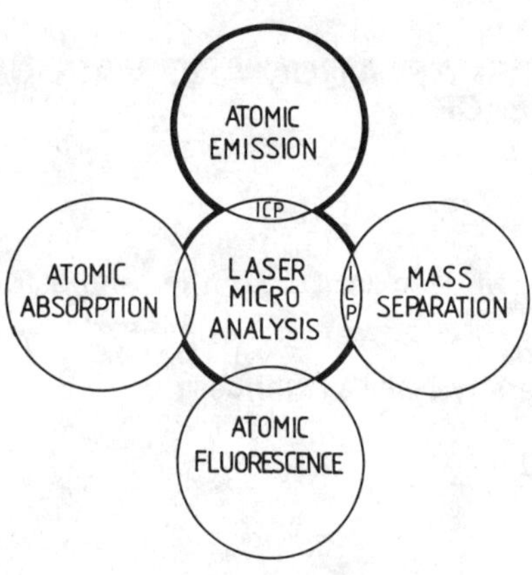

Figure 1 Principles of laser microanalysis

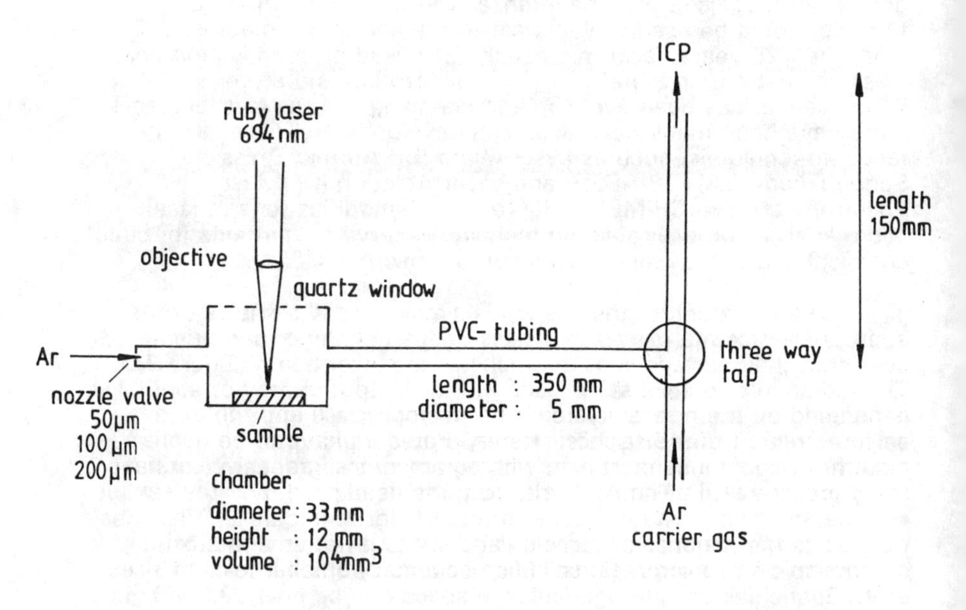

Figure 2 Principles of LA-ICP system

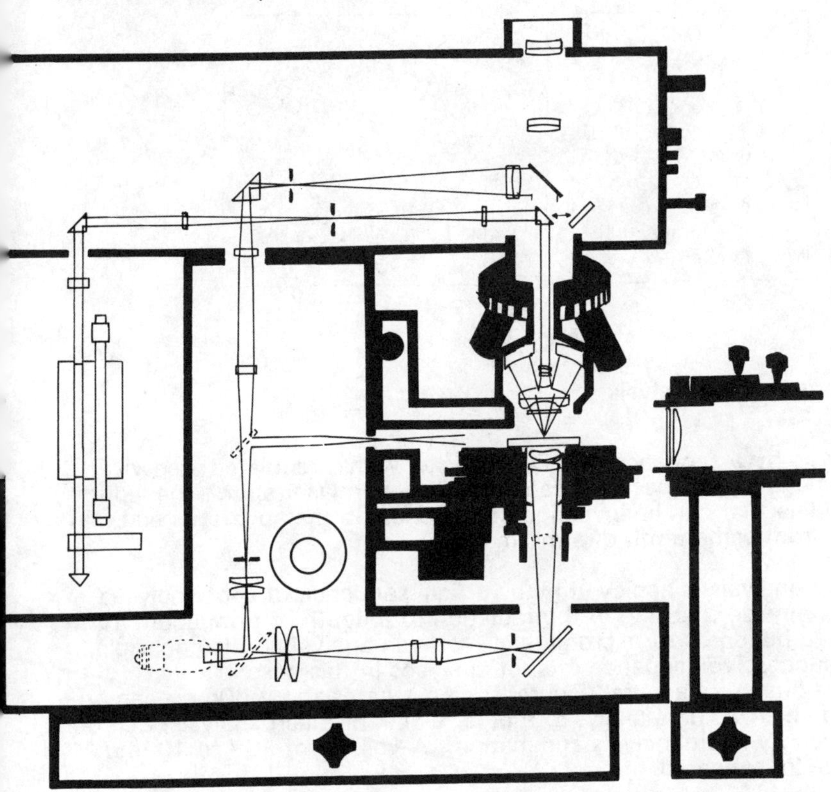

Figure 3 Operation mode of a laser microscope

depending on laser parameters, the focusing optics and the performance of the ICP-spectrometer. Powders have to be applied quantitatively to a support disc with or without an adhesive, or have to be pressed into pellets or fused to glass beads.

Laser microanalysis permits also the qualitative and quantitative determination of microregions. Microscopically small local inhomogeneities or heterogeneities in solids can be analyzed without isolation from their surrounding matrices. There is a variable working range from about ten to some hundreds of micrometers. In addition to these two aims, the method also provides information on the change in composition of a specimen as a function of the spatial coordinates - which means that distribution analysis is possible. A line analysis is

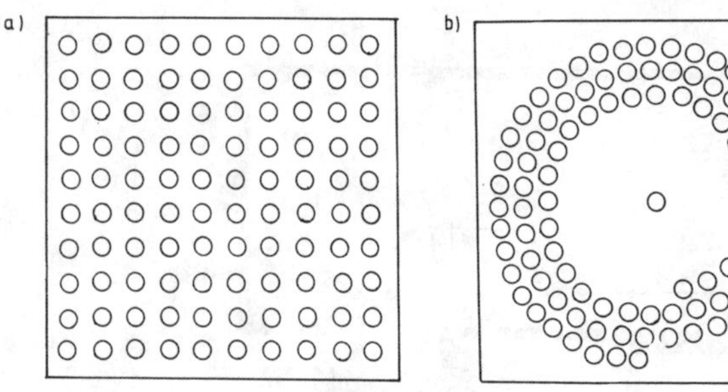

Figure 4 Area analysis

achieved by a sequence of local analyses with a regulated stepwise movement of the specimen carriage after each laser spot. The length of the line that can be investigated depends on the apparatus and may range from some tenth of a millimetre.

A layer analysis is achieved by a vertical sequence of spot analyses after stepwise focusing of the radiation to a depth of from about 1μm to some hundreds of micrometers. An area analysis is also possible, by a successive sequence of spot analyses in this case in x-y dimensions or as a spiral (Figure 4). An area of about 50mm^2 can be recorded. A volume analysis, that is, a bulk or macroanalysis, can be obtained by microanalysis summation. A volume of 10^{-2} to 10^{-1}mm^3 may be investigated.

2 THE LASER AS A RADIATION SOURCE

For laser micro analysis or laser ablation techniques, solid-state lasers in pulsed operation are generally used. Suitable optical media are: ruby-Cr^{3+}doped, glass-Nd^{3+} doped and YAG (Yttrium-Aluminium Garnet)-Nd^{3+} doped. The optical medium is contained in a resonator. In parallel with the resonator, there is a flash-tube which is used as an optical pump. Both units are surrounded by a reflector. If the laser is simply pumped by a pulsed flashtube and the radiation allowed to emerge when the threshold conditions for laser operation are reached, one has what is generally termed a normal mode laser.

Q-switching, changing the Q, or quality of the laser resonant cavity, can be done in a variety of ways. The system in this work is based on a ruby laser using a so-called passive Q-switch, by inserting bleachable dyes between the laser rod and the total reflection prism. Six cells of a glass chamber containing vanadyl-phthalocyanine in

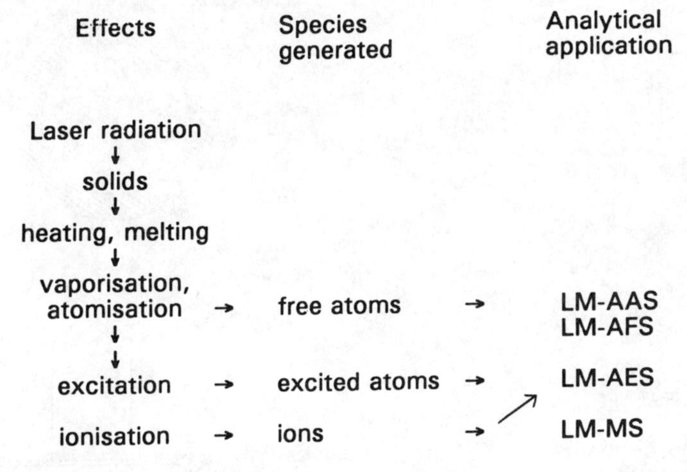

Figure 5 Effects of laser radiation on solids

nitrobenzene may be inserted. With increasing thickness of the absorbing layer, the number of laser pulses falls off until finally only one pulse arises. With normal mode laser pulse widths in the range of 100μs to 1ms are typical. The pulse duration of semi-Q-switched or giant pulses are 100ns to 10μs.

3 LASER - TARGET INTERACTION

Effects caused by absorption of focused laser radiation at the surfaces of solids are heating, melting, vaporisation, excitation and ionisation (Figure 5). Generation of free atoms by vaporisation of solids is the effect used in laser micro analysis based on atomic absorption and fluorescence spectrometry; generation of excited atoms and ions by excitation of the vapour is used in atomic emission and mass spectrometry.

On the presumption that the laser output energy of less than 1 J is turned into heat instantaneously at the point at which the light was absorbed, and absorption, scattering and other effects in the laser plume can be neglected, the total laser output energy has to be equal to the sum, as follows:

$$E_{L\ total} = E_{heat,\ melt,\ vap,\ at,\ ex,\ ion} + E_{losses}$$

Energy losses (Figure 6) by reflection, scattering, thermal

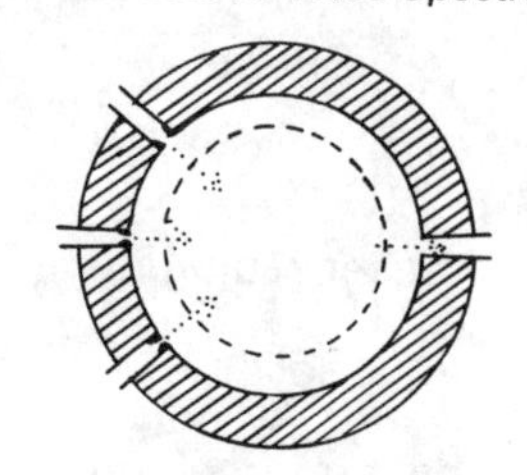

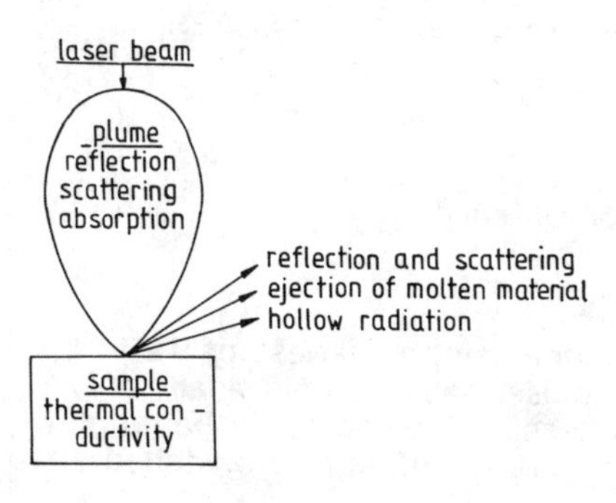

Figure 6 Laser - target interaction

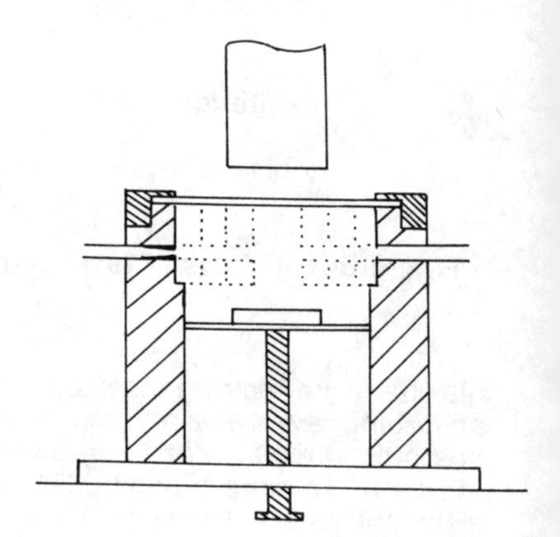

Figure 7 Diagram of sample chamber

conductivity, melting and ejection of both molten and solid material, and by black body radiation are high.

$$E_{losses} = E_{reflected + scattered} + E_{thermal\ condition} + E_{melt} + E_{kinetic} + E_{black\ body\ radiation}$$

The absolute level of energy losses depends on the target material and the mode of laser action.

The energy balance was estimated[1] with the help of physical measurements using steel as the sample material and varying the laser action from normal mode pulses to semi-Q-switched and single Q-switched pulses (Table 1). The reflection and scattering losses depend less on the degree of surface polish and roughness than on the laser mode. In the case of steel a mean value of 2 to 3% must be

Table 1 Energy balance measurements

Mode of laser action	Energy of laser output (J)	Energy losses (% of laser output) E_{r+s}	E_{tc}	E_m	E_k	E_{bbr}	Sum of energy losses (%)	Remaining energy (%)	(J)
Normal mode	0.26	2.0	45	42	0.07	0.9	90	10	0.026
Semi-Q	0.25	2.3	45	33	0.08	0.8	81	19	0.048
	0.22	2.7	45	16	0.12	0.3	64	36	0.079
	0.18	3.0	45	1.6	0.07	0.1	50	50	0.090
	0.10	2.8	45	0.3	0.03	-	48	52	0.052
single pulse	0.06	2.3	45	-	-	-	47	53	0.032

sample = steel

E_{r+s} = $E_{reflected + scattered}$
E_{tc} = $E_{thermal\ condition}$
E_m = E_{melt}
E_k = $E_{kinetic}$
E_{bbr} = $E_{black\ body\ radiation}$
Semi-Q = Semi-Q-switched with decreasing number ofspikes
\- = not determined

considered. The energy lost by thermal conductivity in steel was independent of the laser mode and amounted to 45%. The mass of molten and ejected material was very high in the case of a normal mode laser and decreased with decreasing number of spikes to zero in the case of a single pulsed laser beam. The kinetic energy of the splashes and the black body radiation of the laser crater are less important. By summing up all the energy losses in and above the target the remaining energy has its maximum value in the case of steel in a semi-Q-switched mode. Depending on the kind of target material, an optimisation of laser energy and time duration of the spikes is possible.

Of analytical interest is the threshold condition: the minimum absorbed power density below which no vaporisation will occur (F_{min}), which is described by Ready[4], as

$$F_{min} = d\ L_v\ a^{1/2}\ t_e^{-1/2} \qquad (1)$$

where:

d = mass density of the target
L_v = latent heat of vaporisation of the target
a = thermal diffusivity
t_e = duration of the laser pulse.

Calculating the minimum power density for iron, it follows that it could be vaporised with a normal mode laser of $2x10^6 W.cm^{-2}$. The time predicted to reach the boiling temperature of 3000K is about 4μs. In contrast, with a shorter pulse and a power density of $10^8 W.cm^{-2}$ it needs only 2ns. Such prediction can be calculated by the formula, also given by Ready[4], as

$$t_v = (\pi K d C (T_v - T_o)^2) / 4 F_{min} \qquad (2)$$

Here K, d, C, and F_{min} are the thermal conductivity, mass density, heat capacity per mass unit, minimum power density, and T_v and T_o are the vaporisation and initial temperatures, respectively.

This equation is also useful for rough estimates of surface temperature and depth vaporised. There was a close correlation between signal reproducibility and surface temperature. In the case of metals the mechanism can be described as follows: as more energy (longer pulse duration) is delivered to the surface, the initial vaporisation process changes to the melting-spraying mechanism, by which a significant amount of material is lost. Under optimised conditions the stream of vapour leaves the surface with a velocity of the order of $10^4 cm.s^{-1}$. Q-switched lasers are typified by lower energies, much shorter pulse length, and therefore higher power densities in the range of 10^9 to $10^{12} W.cm^{-2}$. The surface temperature is higher than the boiling temperature and the vapour is partly ionised. The vapour stream leaves the surface with velocities of $10^6 cm.s^{-1}$.

4 TRANSPORT PROCESSES

The vaporised material is entrained in an Ar gas stream and transported by tube to the inductively coupled plasma, the secondary source for subsequent excitation (Figure 2). Since vaporisation, transport, and secondary excitation take place sequentially, each may be independently optimised; secondary excitation of atoms as well as ion formation are insensitive to the physical and chemical state of the ablated material. In an ideal system, the elemental composition of the vapour in the secondary source would mirror that of the sample.

Several chamber designs have been systematically evaluated by others[5-9] and by us for gas flow entrainment. The chamber[10] used in this work has three gas flow entrance nozzle valves of 50, 100 and 200μm and one outlet close to the sample surface (Figure 7).

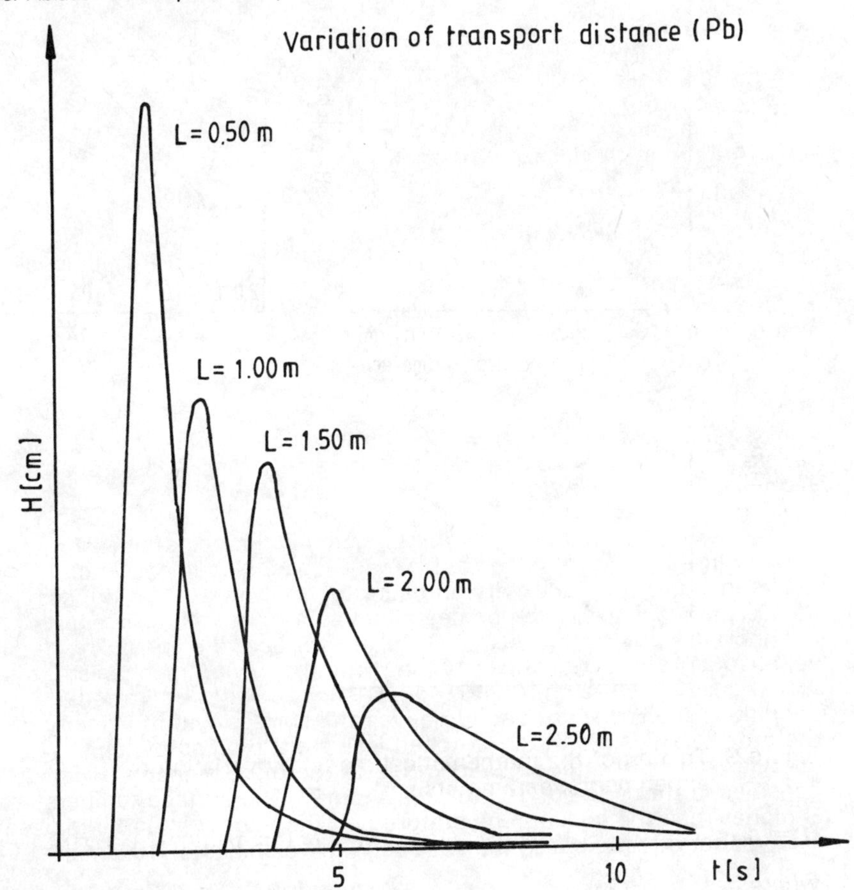

Figure 8 Dependence of emission distribution variable with time on transport distance

Transport gas at atmospheric pressure flows through the chamber with a flow rate of about 0.7L.min^{-1}. The vaporised material is transported through the PVC-tubing of 5mm inner diameter and length 350mm (Figure 2). The height of the signals depends on the length of the tubing, as shown in Figure 8. Analogous results were published by Ishizuka and Uwamino[11]. For theoretical foundation of the transport process we propose to use the formula for dispersion in flow injection analysis, changing some details for our purpose.

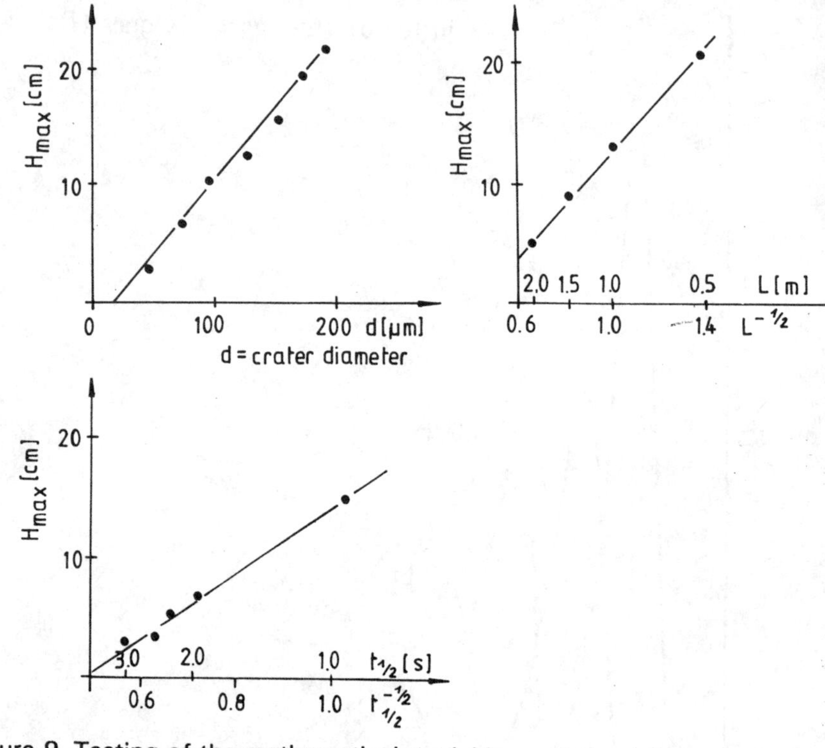

Figure 9 Testing of the mathematical model by variation of experimental parameters

$$D = c_o/c_{max} = (k_1 H_o) / (k_2 H_{max}) = 2 \pi^{3/2} r^2 L^{1/2} V^{1/2} T^{1/2} s_v^{-1} \delta^{1/2} \quad (3)$$

where:

D = dispersion
c_o = injected concentration
c_{max} = concentration corresponding to peak maximum
H_o = signal height of start impulse
H_{max} = peak height
$k_{1,2}$ = factors
r = radius of flow system
L = transport distance
V = flow velocity
T = residence time
s_v = sample volume
δ = dispersion number

and,

$$c \sim k\, H_{max} \sim m / (r^2 \sqrt{L} \sqrt{t_{½}}\, V_{ch}) \qquad (4)$$

where:

c = concentration of the analyte
k = factor
H_{max} = maximum peak height
m = sample mass
r = tube radius
L = tube length
$t_{½}$ = half-with of peak height
V_{ch} = chamber volume

The mathematical model was checked by variation of parameters, such as crater diameter (instead of sample mass), tube length L, and $t_{1/2}$, the half-width of peak height (Figure 9). In all cases proportionality was realised. The influence of transport conditions are represented in Figure 10. Under open and straight tube conditions laminar flow must be considered. For laminar flow, the velocity of gas at the wall tends to zero, while in the centre , the gas velocity is twice the average $v_m = 0.5\, v_{max}$, where v_m is the mean velocity. Trying to use turbulent flow, we can decrease the maximum velocity and increase the mean velocity as shown in the lower part of the Figure 10: $v_m = 0.8...0.9\, v_{max}$. This results a higher sample density in the tube. Turbulent flow can be accomplished by making use of a simple mixing device of a tube packed in a zigzag fashion with glass beads of relatively large diameter (up to 80 % of the inner diameter of the tube): this gives a, so-called single bead string reactor (SBSR)[13], well-known in flow injection analysis (Figure 11). The advantages of SBSR's include the increase of peak heights and the decrease of peak widths (Figure 12), and good reproducibility of the peaks.

For quantification of the signal description, we assume that the sample vapour is ideally mixed in the chamber and that the carrier gas transports the vaporised mass, m, of the sample which is distributed in the chamber volume V_{ch}. Therefore the momentary sample density Q (= f(t)) will be equal to m (= f(t)) divided by V_{ch}.

$$Q = m / V_{ch} \qquad (5)$$

The change of the mass dm is equal to Q F dt, where F is the flow rate in $L.min^{-1}$.

$$dQ = dm / V_{ch}\ ;\ dm = Q\, dv_{gas} = Q\, F\, dt \qquad (6)$$

Therefore we get the simple differential equation:

$$- dQ/Q = (F / V_{ch})dt \qquad (7)$$

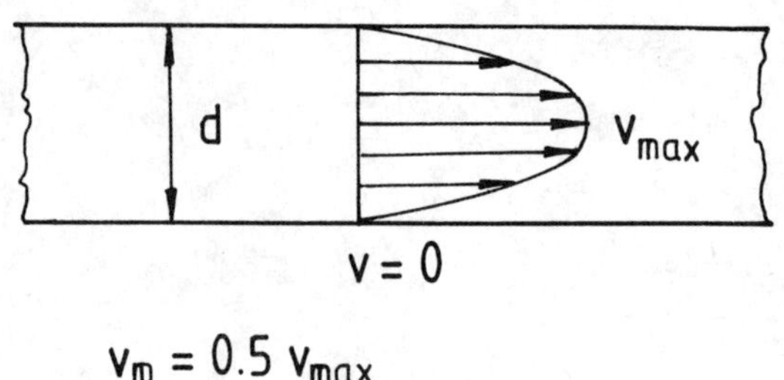

$v_m = 0.5\ v_{max}$

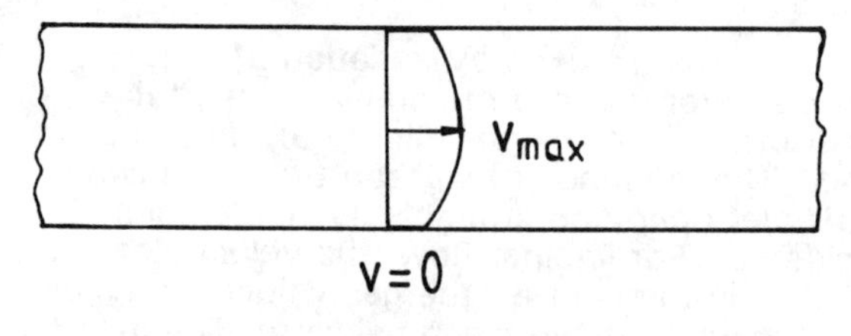

$v_m = 0.8 \ldots 0.9\ v_{max}$

v_m - mean velocity

Figure 10 Velocity profiles in round tubes

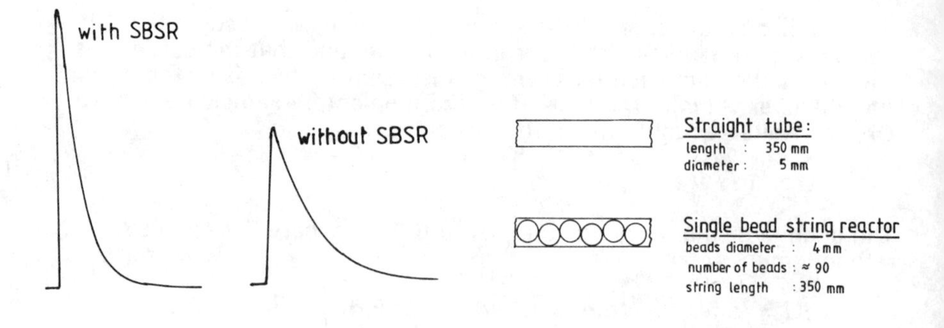

Figure 11 Diagram of open and packed tubes

Figure 12 Peak profiles with and without SBSR

and after integration

$$- \ln Q = (F / V_{ch})t + c \qquad (8)$$

where

$$c = \ln Q^o$$
$$(t \rightarrow 0 ; Q \rightarrow Q^o ; Q^o = m_{total} / V_{ch}),$$

the result:

$$\ln (Q / Q^o) = - (F / V_{ch})t \qquad (9)$$

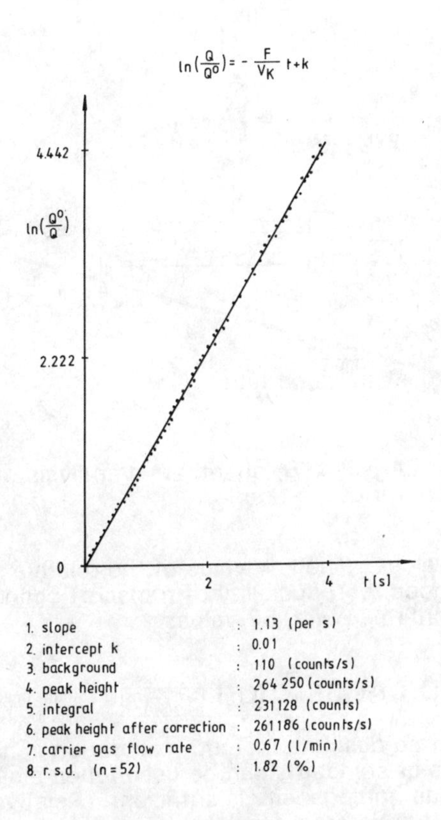

Figure 13 Interpretation of signals obtained by SBSR

After regression with adding a free constant k (Figure 13) it is possible to calculate the integral by

$$I = (H_{max}\ V_{ch}) / F \ , \qquad (10)$$

to correct the peak height using the intercept k by

$$H_{max}{}^{k} = H_{max}\ exp^{k}, \qquad (11)$$

to determine the carrier gas flow rate F using the slope (the chamber volume is known: $V_{ch} = 10^{-2}L$), and to ascertain the relative standard

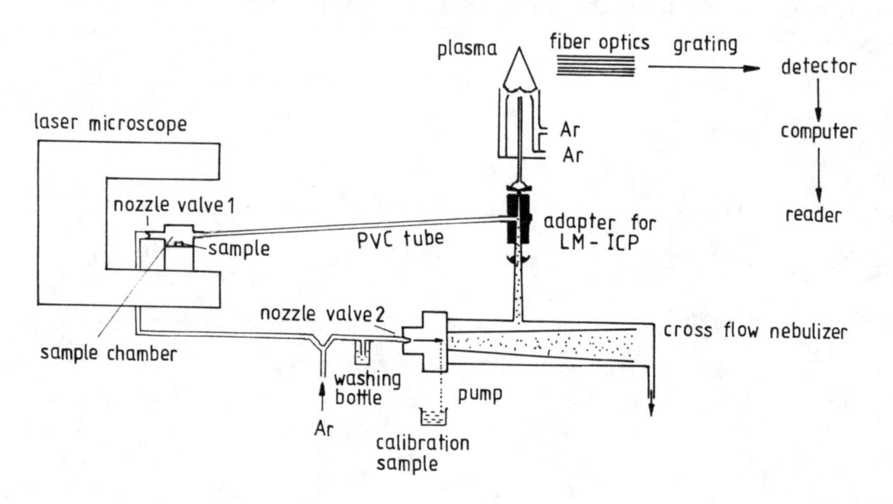

Figure 14 Diagram of LA-ICP-AES; quantitative analysis of solids using liquids for calibration

deviation of measured values. The low value of the relative standard deviation confirms the good reproducibility of transport conditions and exaltation processes, confirmed with 52 values.

5 LIQUID-SOLID CALIBRATION

Since LM-ICP-AES is not an absolute method, calibration is necessary. Solid reference materials or standards can be used when a similar matrix is available and the homogeneity is sufficient. Relative standard deviations are a few % if an internal standard is used and can be about 10% and more if no element is suitable for reference.

In cases where inhomogeneity is troublesome, and that's very common, we are using the method of liquid calibration in a similar manner to Thompson, Chenery and Brett[14]. The experimental equipment of our LM-ICP-Emission Spectrometer uses a parallel flow system of the Ar carrier gas (Figure 14). Both streams can be regulated by nozzle valves. The set-up meets the requirements for analysis of both the solid analyte and the liquid standard under identical spectroscopic conditions, and for simultaneous multielement analysis for two elements at least, the element to be determined and a reference element. In the first step, calibration curves from aqueous solutions were established after nebulising in the normal manner, but carried by one part of the divided Ar stream. The other part of the Ar stream flows through the empty chamber. In the second step, the counted intensities from laser vaporised solids were obtained using both flow streams: one is flowing through the ablation chamber and the other is transporting a blank solution. Both streams are mixed in front of the ICP torch with the aim that the blank introduction into the plasma is the same, thus realising similar plasma conditions.

The desired analyte concentration is calculated by using the formula

$$C_s^a = (C_l^a / C_l^r) C_s^r \qquad (12)$$

where:

C = concentration
a = analyte
r = reference element
s = solid
l = liquid

The relative standard deviation of this calibration method is 3 to 4% [15,16].

If a trace element in a more complex matrix is to be determined, and the contents of all matrix elements are known, the following equations are useful[17]:

$$C_s^a = (C_l^a\ 100\%) / (\Sigma^n_{m=1}\ C_l^m) \quad x) \qquad (13)$$

where:

C = concentration
a = analyte
s = solid
l = liquid
m = matrix elements
x) = valid for simultaneous measurements of all matrix elements

and,

$$C_s^a = (C_l^a / C_l^r) 100\% / \Sigma^n_{m=1} (C_l^m / C_l^r) \quad x) \quad (14)$$

where:

r = reference element
x) = valid for sequential measurement of all matrix elements with simultaneous measurement of one reference element

6 CONCLUSIONS

Laser ablation offers a significant potential for direct solid investigation in micro regions. Effects caused by absorption of focused laser radiation at the surfaces of solids are heating, melting, vaporization, atomisation, excitation, and ionization. The behaviour of surfaces struck by normal laser pulses with millisecond duration, compared to Q-switched laser pulses in the nanosecond region is greatly different.

Energy balance measurements show that the energy losses in and above the target by reflection and scattering of laser radiation, thermal conductivity and ejection of liquid material are high. The absolute height depends on the mode of laser action and on the target material. Therefore, knowledge of the energy situation is important for optimisation of analytical conditions.

To combine laser ablation with inductively coupled plasma excitation, the sample is set up in a chamber. The sample vapour produced from the surface by a laser shot, is introduced to the ICP by the flow of argon carrier gas. The effect of sample chamber design, its volume and aerodynamics, length of tubing and of the transport behaviour of the sample vapour are of interest. It is proposed to use the formula for dispersion of flow injection analysis to interpret the transport processes. Determination of contents in solids can be done by using liquids for calibration after dividing the carrier gas stream in two parts.

ACKNOWLEDGMENT

The authors gratefully acknowledge the instrument support from Spectro Analytical Instruments, Kleve, FRG.

REFERENCES

1. L. Moenke-Blankenburg, 'Laser Microanalysis', Vol. 105 in Chemical Analysis - A Series of Monographs on Analytical Chemistry and its Applications, Eds.: P.J. Elving and

J.D. Winnefordner; I.M. Kolthoff, Ed. em., John Wiley & Sons, New York, 1989.
2. M. Thompson, J.E. Goulter and F. Sieper, *Analyst*, 1981, **106**, 32.
3. A.L. Gray, *Analyst*, 1985, **110**, 551.
4. J.F. Ready, 'Effects of High-Power Laser Radiation', Academic Press, New York, 1971.
5. The VG LASERLAB, VG ELEMENTAL Ltd., Ion Path, Road Three, Winsford, Cheshire CW7 3BX, England.
6. Model 320 Laser Sampler, PERKIN - ELMER Corp., 761 Main Ave., Norwalk, CT 06859-0012 U.S.A.
7. C. Tye, J. Gordon and P. Webb, *Int. Laboratory*, 1987, Dec., 34.
8. P. Arrowsmith and S.K. Hughes, *Applied Spectroscopy*, 1988, **42**, 1231.
9. T. Mochizuki, A. Sakashita, H. Iwata, T. Kagaya, T. Shimamura and P. Blair, *Analytical Sciences*, 1988, **4**, 403.
10. D. Günther, M. Gäckle, J. Kammel, 'Probenkammer zur Erzeugung von Plasmen', WP GO1N 330-747/8, 1989.
11. T. Ishizuka and Y. Uwamino, *Spectrochim. Acta*, 1983, **388**, 519.
12. M. Gäckle and D. Günther, *Z. Chem.*, 1988, **28**, 227 and 258.
13. J.M. Reijn, W.E. Van der Linden and H. Poppe, *Anal. Chim. Acta*, 1981, **123**, 229.
14. M. Thompson, S. Chenery and L. Brett, *J. Anal. At. Spectrom.*, 1989, **4**, 11.
15. D. Günther, M. Gäckle, L. Moenke-Blankenburg, and H.-P. Abicht, *Silikattechnik*, 1990, **41**, 10.
16. L. Moenke-Blankenburg, M. Gäckle, D. Günther, and J. Kammel, Proc. Plasma Spectrometry in the Earth Sciences, Kingston, 1990 (in press).
17. V. Grimm, Diploma thesis, Martin-Luther-University Halle-Wittenberg, GDR, 1989.

INTRODUCTION OF MICROSAMPLES INTO PLASMAS

K. Dittrich, I. Mohamad and K. Niebergall

Karl Marx University
Department of Chemistry
Leipzig DDR 7010
German Democratic Republic

1 INTRODUCTION

For trace analysis in microsamples very sensitive analytical methods are needed, because the quantity to be determined is limited, not only by the content, but also by the sample size. Additional requirements for such methods, are the capability for simultaneous or fast sequential procedures, and for the determination of "all", or at least, most elements, this is because microsamples can be analysed only once by destructive methods.

Mass spectrometric and atomic spectrometric methods, and their combinations have the potential for solving such analytical problems. Highly efficient sample introduction is one of the most important requirements for achieving high absolute sensitivity by the techniques mentioned. In recent years one of the combined, so-called two step procedures, ICP-MS (inductively coupled plasma with mass spectrometric detection) has developed very rapidly. The high degree of ionisation achieved in the high temperature ICP, and the high potential for sensitive and selective detection commonly used quadrupole mass spectrometers, leads to high sensitivity (ng.mL^{-1} to pg.mL^{-1} range).

Sample introduction is normally achieved by nebulisation of the sample solution. The advantage of this sample introduction technique is its stability and reliability. In addition to these, solution nebulisation is simple and straightforward with pneumatic nebulisers. Its disadvantage consists in the low efficiency, which is normally only 1 to 2%. This disadvantage can be improved by other sample introduction methods such as electrothermal vaporization (ETV) for solutions or laser ablation for solids. Table 1 presents an overview of some combined procedures using these more efficient introduction methods. There are one, two and three step procedures. The underlined techniques are those most often used to date. In this paper

Table 1 Sample introduction procedures in atomic and mass spectrometric systems for microsamples

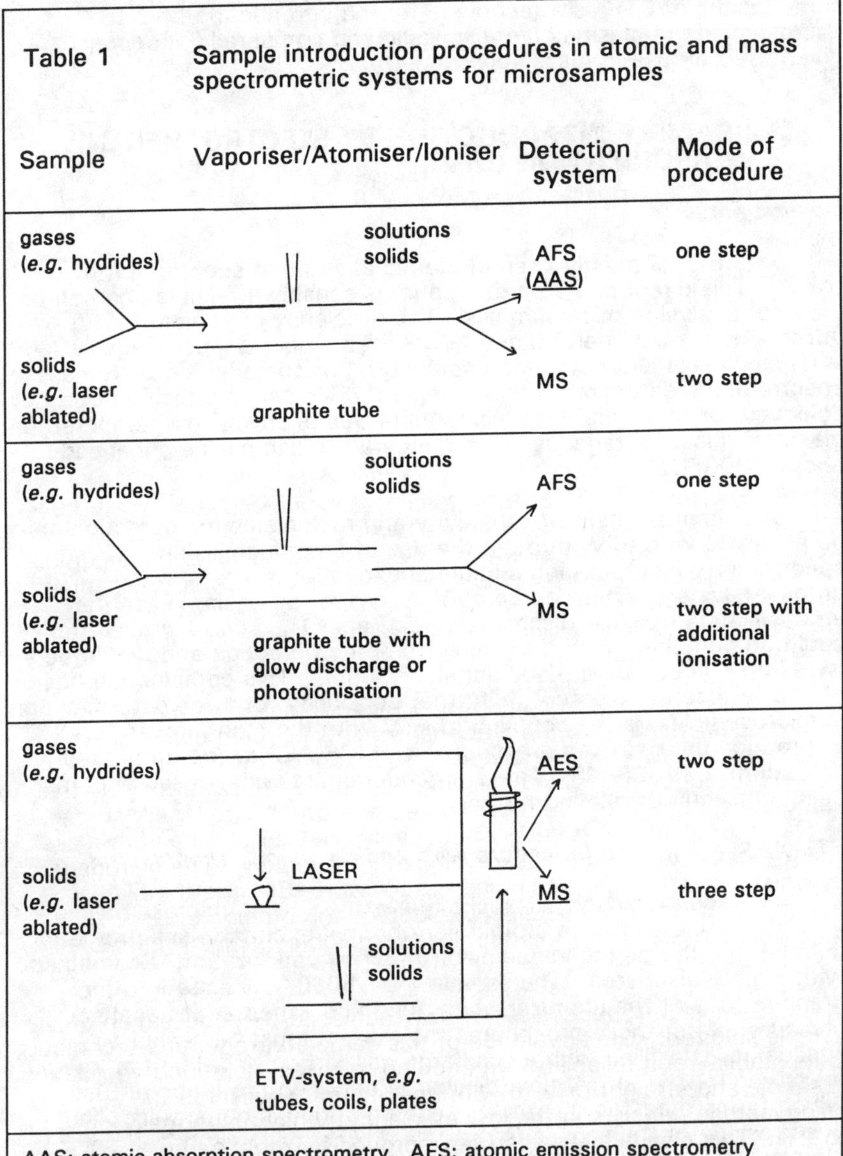

AAS: atomic absorption spectrometry. AES: atomic emission spectrometry
MS: mass spectrometry. ETV: electrothermal vaporisation. LASER: laser ablation

some results of ETV as a technique for the introduction of microsamples to plasmas are discussed and compared. Proposals are given for their use in mass spectrometry.

2 SAMPLE INTRODUCTION BY ELECTROTHERMAL VAPORIZATION (ETV)

General notes

ETV is mostly used in atomic absorption spectrometry (AAS). This direct one-step procedure is sensitive, reliable and can be used for analysing microsamples. The efficiency of sample introduction is between 50 and 100%. The main disadvantage of AAS is its single element detection capability. For combination with mass spectrometric detection, the low degree of ionisation which can be achieved for the majority of elements in this relatively low temperature plasma,is disadvantageous. The temperature can not be enhanced above 3000°C.

Improvement of sensitivity and multielemental detection can be achieved with ETV, through the use of atomic emission spectrometry incorporating additional excitation sources in the atomiser system. This can be carried out, for example, inside the atomiser tube by glow discharges (see Table 1). Such a procedure could be combined with mass spectrometric detection and would be a two-step procedure with additional ionisation. This combination has not so far been attempted. A further possibility for trace determination in microsamples, is the coupling of ETV with the high temperature ICP to produce the two-step procedure, ETV-ICP-AES,or the three-step procedure, ETV-ICP-MS. These procedures are widely used and are being currently developed further.[1-4]

ETV-AES: a one-step procedure with additional ionisation/excitation

A recently developed technique of AES using microsampling in graphite tubes is furnace atomic non-thermal excitation spectrometry (FANES)[5-7]. In this technique electrothermal vaporization is combined with a glow discharge. The sample (10 - 50μL) is loaded into the graphite tube of the atomizer, then dried and ashed at atmospheric pressure. After these procedures the system must be sealed, evacuated, and filled with a discharge gas (argon or helium) at low pressure (1 - 3 kPa). A glow discharge between the graphite tube - which forms the hollow cathode - and an anode is generated. The dried residue of the sample is then electrothermally vaporized and atomised by fast heating of the tube and the glow discharge. The free atoms generated are excited and partly ionised in the discharge.[8]

Figure 1 shows that at higher temperature there are ions which could be used in principle for mass spectrometric detection. If helium

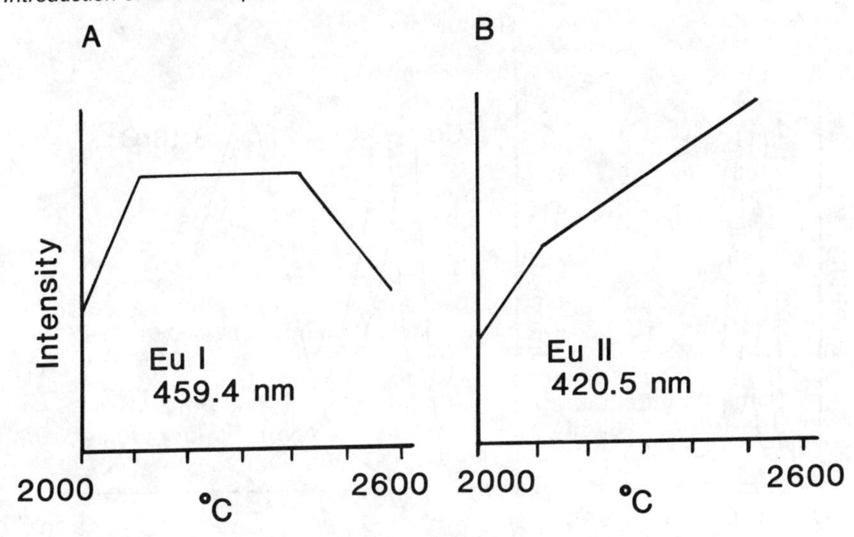

Figure 1 Dependence of Eu - FANES (a) and Eu^+ - FINES (b) on the atomisation temperature
Eu - amount: 2 μg.mL^{-1}
EuO → Eu + O (esp. 2000° - 2100°C)
Eu → Eu^+ (esp. at higher temp.)
FINES - Furnace ionisation non-thermal excitation spectrometry

is used instead of argon the degree of ionisation can be greatly enhanced, particularly for elements of high ionisation energies, such as the non-metals[5]. For combination with mass spectrometric detection, the low pressure in the FANES source and the possible close coupling to the mass spectrometer may be advantageous.

ETV-ICP AES: a two-step procedure

ETV-ICP-AES and (-MS) procedures use metal element atomizers[1-3] and carbon plate atomizers.[4] For our study, a new tungsten coil atomizer[9] and the graphite tube atomizer of the FANES source described above are used. The closed character of the FANES source is advantageous for the combination with an ICP system, when this source is used only as vaporizer. In following the closed character, there is no loss of substance through the dosing hole, entrance of air is avoided, and the gas flow can be easily optimised.

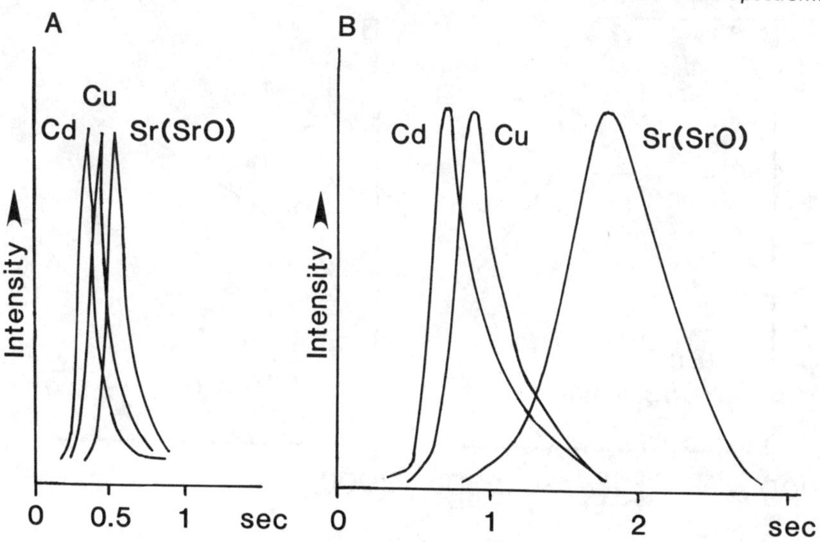

Figure 2 Intensity - time curves of ETV-ICP-AES using tungsten coil (a) and graphite tube FANES - atomiser (b) for different elements

heating rates (a) 10K.ms^{-1}
(b) 2K.ms^{-1}
boiling points Cd - 765°C
Cu - 2595°C
SrO - 2800°C
note: Sr evaporates as SrO/Sr - carbide (b) and SrO (a)

Especially in AES detection, short emission signals are needed for high signal to background ratios. Figure 2 shows the distinction in heating rate and signal formation for the tungsten coil and the carbon tube atomiser. In the case of the tungsten coil atomiser the signals of all elements measured are generated during the first second of vaporization. This is caused by the high heating rate exceeding 10K.ms^{-1}. In the case of the carbon tube the heating rate is only 2K.ms^{-1}. Therefore a longer integration time for multielemental analysis (3s) is needed. The half width of the signals produced with the tungsten coil atomiser are smaller than those of the carbon-tube atomiser. The`difference is not very large.

Table 2 Absolute detection limits (pg, 3σ) of ICP-AES using nebulisation, ETV (in tungsten coil and graphite tube atomiser) and FANES.

Element	Boiling point (°C)	Nebulisation	W-ETV	C-ETV	FANES
Cd	765	34,000	90	-	0.4
Zn	907	145,000	1,700	60	2
Ag	2212	12,500	7	24	0.4
Cu	2595	12,000	11	20	1
Yb	1194*	40,000	18	-	20
Sr	1384*	500	-	0.5	0.02
Lu	3395*	14,000	54	-	10,800

* carbide forming elements
sample volumes: 20μL - ETV, FANES
100μL - nebulisation

Comparison of analytical results

In Table 2 some analytical results are shown for different analytical procedures. The following deductions may be made from these results.

Firstly, the absolute detection limits of nebulisation ICP-AES can be improved very greatly by either form of ETV-ICP-AES procedures (by about 3 orders of magnitude). Considering the different sample volumes needed (nebulisation *versus* ETV) the improvement of the relative detection limits (related to the volume of the sample solution) is only one order of magnitude.

Secondly, the absolute detection limits obtained in ETV-ICP-AES are of the same order of magnitude for both the tungsten coil atomiser and the carbon-tube FANES-ETV system for the elements measured. The detection limit of zinc is much worse in the tungsten coil atomiser than from the graphite system, as a result of a high background due to zinc contamination of the tungsten. A special problem of the tungsten coil ETV-ICP-AES consists in the possible line coincidences by the line rich element tungsten, but this is not a problem in ETV-ICP-MS.

Thirdly, comparing the analytical values of both two-step ETV-ICP-AES procedures with the one-step FANES technique, it can be seen that the absolute detection limits of the FANES are 1-2 orders of magnitude better than those of ETV ICP-AES for easily non vaporizable, non-carbide forming elements. Refractory, carbide-forming elements can be determined most sensitively with the tungsten coil ETV-ICP-AES[9].

3 CONCLUSIONS

By the combination of ETV with ICP-AES and ICP-MS techniques, sample introduction efficiency,and as a result the absolute detection limits, have been dramatically improved for microsamples, in comparison to normal nebulisation techniques.

The one step procedure, FANES, gives much better detection limits for easily and moderately vaporizable elements than the ETV ICP-AES techniques. The reason for this is the direct vaporization of the sample into the plasma. In addition, when compared to ETV-AAS, FANES also offers a strong enhancement due to the degree of ionisation in this source. Therefore, it can be recommended to test this technique also in combination with MS detection. An additional advantage of this set-up should lie in the vaporization, excitation and ionisation under vacuum conditions.

Acknowledgement: The authors wish to thank the Company Spectro Analytical Instruments, Kleve F.R.G., for some help in instrumentation.

REFERENCES

1. D.F. Nixon, V.A. Fassel and R.N. Kniseley, *Anal. Chem.*, 1974, **46**, 210.
2. C.J. Park, J.C. van Loon, P. Arrowsmith and J.P. French, *Anal. Chem.*,1987, **59**, 2191.
3. C.J. Park, J.C. van Loon, P. Arrowsmith and J.P. French, *Can. J. Spectrosc.*, 1987, **32**, 29.
4. G.E.M. Hall, J.C.L. Pelchat, D.W. Boomer and M.Powell, *J. Anal. At. Spectrom.*, 1988, **3**, 791.
5. H. Falk, E. Hoffmann and Ch. Lüdke, *Progr. Anal. Spectrom.*, 1988, **11**, 417.
6. H. Falk, E. Hoffmann and Ch. Lüdke, *Spectrochim. Acta B*, 1981, **36**, 767.
7. K. Dittrich and H. Fughs, *J. Anal. At. Spectrom.* 1987, **2**, 533.
8. K. Dittrich, G. Rismann and H. Fughs, *J. Anal. At. Spectrom.*, 1988, **3**, 459.
9. K. Dittrich, H. Berndt, J.A.C. Broekaert,G.Schaldach and G. Tölg, *J. Anal. At. Spectrom.* 1988, **3**, 1105.

INVESTIGATIONS ON MIXED GAS PLASMAS PRODUCED USING A SHEATHING DEVICE IN ICP-MS

Diane Beauchemin and Jane M. Craig

Department of Chemistry,
Queen's University
Kingston, Ontario,
Canada K7L 3N6

1 INTRODUCTION

Inductively coupled plasma mass spectrometry (ICP-MS) extends the sensitivity and selectivity of mass spectrometry to the multielemental analysis of solutions[1]. It features low detection limits, relatively simple spectra and the capability of isotopic analysis[1]. The technique does however have a number of limitations, principally due to occurrence of isobaric overlaps and polyatomic ion interferences which can arise from both the background and the matrix[2,3]. For some elements an alternative isotope, free from interference, is available, but some elements are monoisotopic and no alternative can therefore be used, giving rise to a degraded performance. Most ICP-MS instruments use an argon plasma which produces background peaks directly overlapping the major isotopes of K, Ca, Fe (ArO) and Se (ArAr), dramatically degrading the detection limits for these elements. Multivariate calibration can, to some extent, be successful in circumventing these interferences[4]. However, it requires a large number of standards, and samples must be judged on a case-by-case basis since the interferences must first be identified in order to select the simplest calibration scheme. Gas-phase collisions can also be used to attenuate interferences but this requires either a double[5] or a triple[6] quadrupole arrangement where one of the quadrupoles is used in an rf-only mode, within a collision cell. Means to prevent the formation of polyatomic ion interferences would solve a major part of the problem, directly at its source. Interferences due to Ar alone would remain.

A few workers have reported that the presence of another gas in the argon changes the fundamental properties of the plasma[7-9]. For instance, the addition of a few percent of hydrogen as a sheath around the nebulizer gas flow enhanced the ionization process in inductively coupled plasma atomic emission spectrometry (ICP-AES)[7]. This has been attributed to the higher thermal conductivity of these diatomic gases, which leads to a more efficient transfer of energy within the plasma, resulting in improved desolvation, volatilization and dissociation processes[7]. In ICP-MS, the addition of a diatomic gas to the plasma gas was found to improve the analytical performance when the operating conditions were modified[8]. For instance, the addition of 5% nitrogen to the plasma gas resulted in an increase in sensitivity when higher nebulizer flow rate and RF power, and smaller sampling depth were used[8]. A reduction of the interferences from $ArCl^+$ and Ar^{2+} on arsenic and selenium at masses 75-78, was also reported to result from the addition of a small amount of oxygen or nitrogen to the nebulizer gas[9]. The addition of a small amount of propanol had a similar effect[9].

The purpose of this study was to investigate the use of mixed gas plasmas in an attempt to find operating conditions that would fulfil at least one of the following goals: (i) improve the analytical characteristics of ICP-MS (especially for Fe and Se), (ii) allow a reduction in the nebulizer flow rate (to enable the coupling of ICP-MS with techniques requiring a carrier flow rate smaller than 1L.min^{-1}, such as gas chromatography (GC)). The mixed gas plasmas considered were produced by adding a sheathing gas around the nebulizer gas (Figure 1).
The effect of argon was first assessed (as a reference) and compared

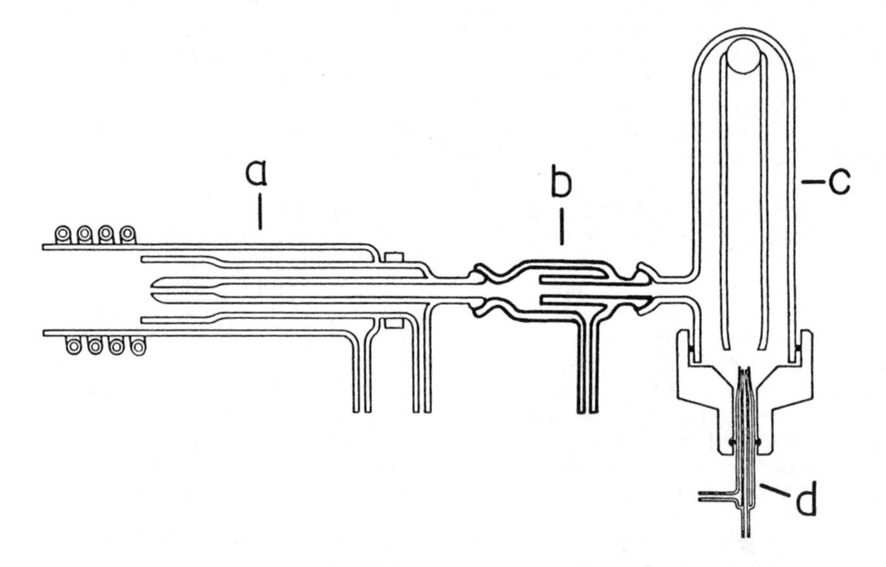

Figure 1 Schematic of the set-up used for the addition of a sheathing gas. (a) low-flow PlasmaTherm torch; (b) sheathing device; (c) double-pass spray chamber; (e) Meinhard concentric nebuliser.

to that of hydrogen and nitrogen. The latter two gases were selected because they were reported to have the most drastic effect on the excitation conditions in the central channel of the plasma in ICP-AES[10].

2 EXPERIMENTAL

Instrumental conditions

The ICP-MS instrument was the Perkin-Elmer SCIEX ELAN 500 (Thornhill, Ontario, Canada). Some modifications were made to the originally supplied instrument. A conventional ICP-AES torch was used instead of the approximately 15 mm longer one that was provided with the instrument. A mass flow controller (Model 1259B, MKS Instruments, Andover, Massachusetts, USA) was installed on the nebulizer gas line. A peristaltic pump (Minipuls II, Gilson Medical Electronics Inc., Middleton, WI, USA) was put on the sample delivery tube in order to ensure a constant sample introduction rate. The addition of a sheathing gas was achieved using a glass sheathing device (Jobin-Yvon, Longjumeau, France) which was installed between the spray chamber and the torch (see Figure 1). The flowmeter originally used to control the nebulizer gas, now supplied by a mass

Table 1. ICP-MS operating conditions.

Plasma conditions	
torch	low flow (PlasmaTherm)
RF power	1.2 - 1.4 kW
reflected power	≤5 W
plasma gas flow	9.0 - 10.0 $L.min^{-1}$
auxiliary gas flow	2.0 $L.min^{-1}$
nebulizer gas flow	0.08 - 0.83 $L.min^{-1}$
sheath gas flow	0 - 1.0 $L.min^{-1}$
sample delivery rate	1.0 $mL.min^{-1}$
Mass spectrometer settings	
Bessel box stop	-6.1 V
Bessel box barrel	2.0 V
Einzel lenses 1 & 3	-12.9 to -16.3 V
Bessel box end lenses	-6.2 V
Pt sampler orifice diameter	1.14 mm
Pt skimmer orifice diameter	0.89 mm
interface pressure	0.8 - 0.9 Torr
mass spectrometer pressure	(0.8 - 5.0) x 10^{-5} Torr

flow controller, was used to supply the sheathing gas. The operating conditions used throughout this work are summarized in Table 1.

The measurements were made, using the "Multiple Elements" software provided with the instrument, by peak hopping rapidly from one mass to the other, with a dwell time of 10 ms per mass per scan, scanning repeatedly until the total measuring time (per measurement) of 0.5 s was reached. Three measurements were made per peak, one measurement being done at the central mass while the two others were done at ± 0.1 amu from the assumed peak centre. A resolution of 1.1 amu (peak width) at 10% peak height was maintained throughout the studies. The species monitored were: $^{51}V^{+}$, $^{52}Cr^{+}$, $^{55}Mn^{+}$, $^{56}Fe^{+}$, $^{57}Fe^{+}$, $^{60}Ni^{+}$, $^{63}Cu^{+}$, $^{66}Zn^{+}$, $^{76}Se^{+}$, $^{114}Cd^{+}$ and $^{208}Pb^{+}$.

Reagents

A 100 ng.mL^{-1} multielement standard solution was prepared from 1000 μg.mL^{-1} stock monoelemental solutions (Spex Industries, Edison, New Jersey, U.S.A.) for ICP spectrometry. Dilutions were carried out using deionized distilled water (Milli-Q Plus, Millipore, Mississauga, Ontario, Canada) and with the appropriate amount of high purity nitric acid (Seastar, Sidney, British Columbia, Canada) to reach a final concentration of 0.1 M HNO_3.

Analytical procedure

Instrument operating conditions. Both the ion lens voltages of the instrument and the alignment of the plasma with respect to the interface were optimized for an all argon plasma, with a plasma gas flow rate of 9 L.min^{-1} and an RF power of 1.2 kW, without introducing any gas (including argon) through the sheathing device. The optimization was carried out while aspirating a 100 ng.mL^{-1} standard solution containing Li, Rh and Pb. First, the torch positioning, with respect to the sampler, and the nebulizer gas flow rate were adjusted to give maximum sensitivity for $^{103}Rh^{+}$. Then the ion lenses voltages were adjusted such that the $^{103}Rh^{+}$ stayed at a maximum while the $^{7}Li^{+}$ and $^{208}Pb^{+}$ signals were made equal. A measurement time of 0.2 s was used for the optimization process.

Sheathing gas study. The sheathing gas was introduced into an all argon plasma by slowly increasing its flow rate up to the desired value. For the investigations using nitrogen and hydrogen, the plasma gas flow rate was increased to 10 L.min^{-1} and the RF power was set at 1.4 kW before introduction of the sheathing gas. In each case, a qualitative study was first carried out while aspirating a 100 ng.mL^{-1} multielemental standard solution. The nebulizer gas flow rate was optimized for maximum sensitivity while continuously monitoring the signals from $^{51}V^{+}$, $^{52}Cr^{+}$, $^{55}Mn^{+}$, and $^{56}Fe^{+}$. This optimization was carried out for each flow rate and type of sheathing gas. A more detailed study was then carried out at selected sheathing gas flow rates, with the corresponding optimum nebulizer flow rate which was predetermined during the qualitative study. In each case, the blank (0.1 M HNO_3) was aspirated, followed by the 100 $ng.mL^{-1}$ standard solution. Ten determinations were done on the blank and five on the standard. The range of RF power, and plasma gas, nebulizer gas and sheathing gas flow rates covered are listed in Table 1.

The approach used here is different than that taken for ICP-AES[14] where the nebulizer flow rate was not changed (in order to ensure a constant nebulization efficiency) but variation of the observation height was accomplished. Such profiling is currently carried out in ICP-AES, but is difficult to accomplish with the present ICP-MS instrument, because the torch box cannot be positioned precisely and reproducibly with respect to the sampling interface. Therefore, the sampling depth was kept constant by adjusting the nebulizer flow rate (for each sheathing condition).

Data treatment. All standard intensities were blank-subtracted. For each set of conditions, the sensitivity and detection limit for each analyte were measured. The detection limit was defined as the concentration of the analyte needed to give a signal equivalent to three times the standard deviation of the blank. To facilitate comparison, the sensitivities and detection limits for the analytes were also calculated as relative signal percentages of those obtained under usual operating conditions without any sheathing gas. The molar sensitivity for each analyte, corrected for 100% isotopic abundance[11], was also computed.

3 RESULTS AND DISCUSSION

Effect of argon

The sensitivities and detection limits obtained when argon was used as the sheathing gas at the rates of 0.5 and 1.0 $L.min^{-1}$ are compared, in Table 2, to those obtained in usual operating conditions free of sheathing gas (the effects of other sheath gases are shown in Table 3). The results are also shown graphically in Figures 2 and 3, as signal (sensitivity or detection limit) percentages relative to that in usual conditions free of sheathing gas. The optimum nebulizer gas flow rate found in each case indicates that as the sheathing gas flow rate was increased, the nebulizer gas flow rate had to be decreased in order to maximize sensitivity. For 0.5 $L.min^{-1}$ of argon sheath, the best sensitivities were obtained at a nebulizer gas flow rate of 0.45 $L.min^{-1}$, whereas 0.81 $L.min^{-1}$ was the nebulizer gas flow rate required under usual operating conditions. When the argon sheathing flow rate was increased to 1.0 $L.min^{-1}$, the nebulizer gas flow rate had to be decreased further to 0.08 $L.min^{-1}$. The addition of a sheathing gas moves the initial radiation zone away from the load coil (*i.e.* closer to the sampler) and a lower nebulizer flow rate is required to bring it back

Table 2. Sensitivities (counts.s^{-1} per ng.mL^{-1}) and detection limits (ng.mL^{-1}, based on 3σ) observed with a sheath of argon

Analyte	No sheath[1,2]		0.5 L.min^{-1} Ar[1,3]		1.0 L.min^{-1} Ar[1,4]	
ion	sensitivity	DL	sensitivity	DL	sensitivity	DL
^{51}V	1650 ± 18	0.0016	847 ± 11	0.007	2.82 ± 0.28	12
^{52}Cr	1540 ± 20	0.040	777 ± 14	0.095	2.12 ± 0.05	25
^{55}Mn	2370 ± 32	0.044	1200 ± 16	0.046	3.43 ± 0.45	6.7
^{56}Fe	1980 ± 41	4.4	1030 ± 25	12	n.d.	n.d.
^{57}Fe	47.0 ± 1.2	2.0	24.4 ± 1.1	6.4	0.06 ± 0.03	160
^{60}Ni	320 ± 5	0.61	171 ± 2	0.12	0.51 ± 0.08	12
^{63}Cu	667 ± 10	0.057	365 ± 5	0.16	1.14 ± 0.17	15
^{65}Zn	142 ± 2	0.23	81.7 ± 0.8	0.35	0.45 ± 0.04	10
^{76}Se	18.2 ± 0.6	11	10.5 ± 0.4	35	0.39 ± 0.44	450
^{114}Cd	573 ± 7	0.056	280 ± 2	0.015	1.06 ± 0.11	20
^{208}Pb	1160 ± 24	0.092	482 ± 8	0.048	1.62 ± 0.10	27

1 RF power: 1.2 kW; plasma gas flow rate: 9 L.min^{-1}.
2 Nebulizer flow rate: 0.81 L.min^{-1}.
3 Nebulizer flow rate: 0.45 L.min^{-1}.
4 Nebulizer flow rate: 0.08 L.min^{-1}.

to the optimal sampling depth of about 10 mm (measured while aspirating a 500 μg.mL^{-1} Y solution, as the distance between the initial radiation zone[12] and the tip of the sampler).

This observation is in agreement with the results of Lichte and coworkers[13] who reported an increase in count rates with a higher argon sheathing gas flow rate (at 1.6 kW). However, it differs from the behaviour found in ICP-AES by Murillo and Mermet[14] who reported no shift of peak position (*i.e.* observation height) for ionic lines when 0.25 L.min^{-1} of argon was added as a sheath. This suggests that the sampling process in ICP-MS has a non-negligible effect on the processes occurring in the plasma (whereas the measurement of light in ICP-AES is a totally passive process). The total argon flow traversing the injector increased from 0.81 L.min^{-1} with no sheathing gas up to 1.08 L.min^{-1} with 1 L.min^{-1} of Ar. This can probably be attributed to the reduced nebulization efficiency. As the nebulizer flow rate is decreased, the amount of aerosol reaching the plasma is also reduced which increases the sampling depth. A higher nebulizer flow rate is therefore required to compensate for the lower nebulization efficiency and push the initial radiation zone back to the optimal position.

As can be seen from Figure 2, all of the sensitivities obtained

Table 3. Sensitivities (counts.s^{-1} per ng.mL^{-1}) and detection limits (ng.mL^{-1}, based on 3σ) observed with a sheath of hydrogen or nitrogen

Analyte ion	Trace of N_2[1,2] sensitivity	DL	Trace of H_2[1,3] sensitivity	DL	0.05 L.min^{-1} H_2[1,4] sensitivity	DL
$^{51}V^+$	1060 ± 15	0.023	1290 ± 31	0.019	384 ± 5	0.22
$^{52}Cr^+$	889 ± 13	0.33	1110 ± 28	0.056	318 ± 3	0.15
$^{55}Mn^+$	1140 ± 15	1.3	1450 ± 46	0.029	395 ± 4	0.39
$^{56}Fe^+$	946 ± 32	16	1290 ± 56	8.6	358 ± 6	15
$^{57}Fe^+$	23.3 ± 0.3	1.8	30.5 ± 1.1	4.5	8.64 ± 0.19	15
$^{60}Ni^+$	188 ± 3	0.12	239 ± 7	14	77.2 ± 0.9	0.25
$^{63}Cu^+$	429 ± 8	0.24	519 ± 14	0.13	167 ± 2	0.083
$^{66}Zn^+$	136 ± 3	0.25	125 ± 4	2.9	48.8 ± 0.9	0.67
$^{76}Se^+$	13.6 ± 1.2	54	14.2 ± 0.4	7.5	4.80 ± 0.40	51
$^{114}Cd^+$	262 ± 5	0.099	298 ± 11	2.5	91.0 ± 1.2	0.64
$^{208}Pb^+$	272 ± 4	0.11	326 ± 17	0.12	80.9 ± 1.2	0.089

1 RF power: 1.4 kW; plasma gas flow rate: 10 L.min^{-1}.
2 Nebulizer flow rate: 0.6 L.min^{-1}.
3 Nebulizer flow rate: 0.75 L.min^{-1}.
4 Nebulizer flow rate: 0.65 L.min^{-1}.

using 0.5 L.min^{-1} argon sheathing, are lower than those under usual operating conditions. This was to be expected with a lower nebulizer flow rate since the efficiency of the nebulization process was then degraded (*i.e.* less solution was converted into a usable aerosol and/or transported to the torch). It is also in agreement with results from Murillo and Mermet[14] showing a decrease in the ionic lines intensity with the addition of an argon sheathing gas. Based on the detection limits reported for several ionic lines in ICP-AES[7] with and without an argon sheathing gas, one could also expect a concurrent degradation of the detection limits. Figure 2 shows that lower detection limits were the case for most of the isotopes, except $^{55}Mn^+$, $^{60}Ni^+$, $^{114}Cd^+$ and $^{208}Pb^+$, where $^{55}Mn^+$ had essentially the same detection limit whereas the others exhibited significantly improved detection limits. This was especially so for $^{60}Ni^+$ which had a detection limit five times better than that obtained under usual operating conditions.

The improved detection limits are due to a decrease in background count rate, an enhanced signal-to-background ratio (S/B), and/or a more stable background signal due to a more stable plasma. Tables 4 and 5 report the background count rates observed in the various sheathing gas conditions. A comparison of the background for 0.5 L.min^{-1} of Ar sheath to the corresponding one without a sheath shows that a significantly reduced background is observed for all four species with either no change (for $^{55}Mn^+$) or a significant improvement of the corresponding relative standard deviation (RSD) for $^{60}Ni^+$, $^{114}Cd^+$ and $^{208}Pb^+$. In the case of $^{63}Cu^+$ and $^{66}Zn^+$, a decrease in background corresponded to an increase in RSD. No significant change in background was observed for $^{51}V^+$ and $^{57}Fe^+$ (although the RSD was worse) whereas an increase in background was obtained for $^{56}Fe^+$, as well as $^{52}Cr^+$ and $^{76}Se^+$. The background increase at amu 76 is due to a greater amount of argon dimer being formed from the plasma as the argon sheathing flow rate increases.

Table 4. Background count rates (counts s^{-1} ± % relative standard deviation) observed for 0.1 M HNO_3 with and without a sheath of argon.

Analyte ion	No sheath		Ar 0.5 $L.min^{-1}$		Ar 1.0 $L.min^{-1}$	
	counts	RSD	counts	RSD	counts	RSD
^{51}V	17.6	5	16.2	13	9	120
^{52}Cr	687	3	933	3	1157	2
^{55}Mn	786	4	620	3	43	2
^{56}Fe	112900	3	134900	3	53430	1
^{57}Fe	1682	2	1634	3	58.6	6
^{60}Ni	268	24	115	6	3	67
^{63}Cu	565	2	359	5	15.4	36
^{66}Zn	487	2	269	4	4.9	32
^{76}Se	2427	3	3910	3	6278	1
^{114}Cd	74	14	19.4	7	20.7	35
^{208}Pb	438	8	144	5	44	33

Improved S/B were only observed for $^{60}Ni^+$, $^{114}Cd^+$ and $^{208}Pb^+$ (Table 6), *i.e.* those isotopes giving rise to improved detection limits. $^{63}Cu^+$ and $^{66}Zn^+$ did not exhibit a significant change in S/B whereas all other species suffered a decrease, especially $^{76}Se^+$ whose S/B dropped 25-fold. Overall, similar or improved detection limits were obtained for those isotopes whose background signal dropped while either remaining as stable or becoming more stable.

The flow rate of the Ar sheathing gas was also increased up to 1.0 $L.min^{-1}$ to check if a signal could still be detected at a very low nebulizer flow rate. The sensitivities and detection limits obtained are included in Table 2 and shown graphically in Figure 3. The sensitivities are all less than 2% of those obtained under usual operating conditions. This loss of sensitivity is certainly due in great part to a decreased nebulization efficiency experienced upon the 10-fold reduction of the nebulizer gas flow rate. The relative detection limits are all less than 6%, but as with an Ar sheathing of 0.5 L.min^{-1}, the highest relative detection limit observed was for $^{60}Ni^{+}$. In all cases, the S/B was decreased (Tables 4 and 5). The same was true for the background (Table 6), except for $^{52}Cr^{+}$ and $^{76}Se^{+}$ which exhibited increased background but with an improved RSD.

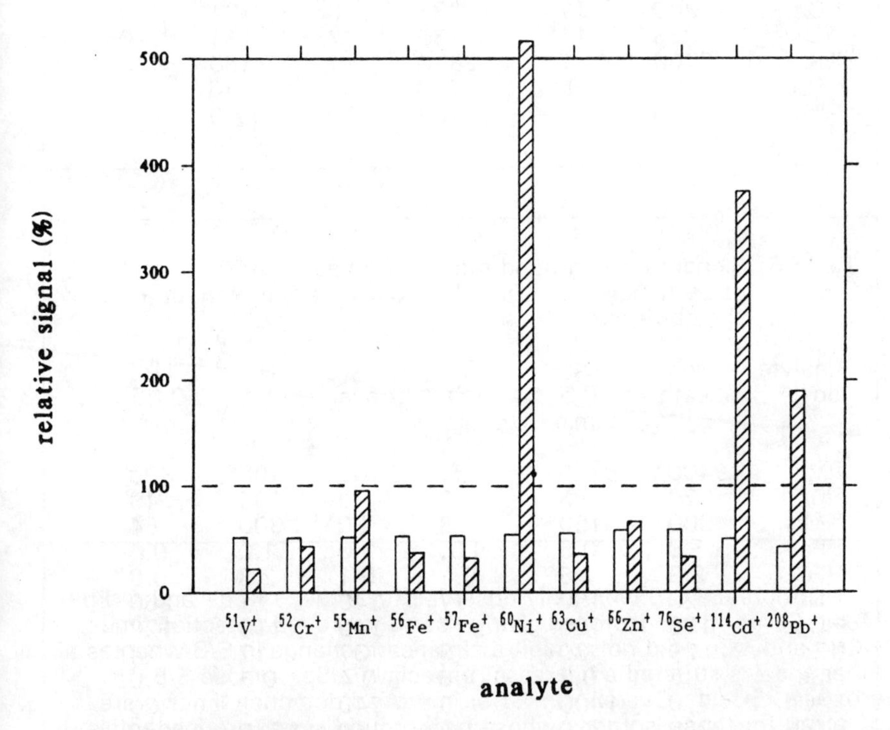

Figure 2 Effect of the addition of 0.5 L.min^{-1} Ar as a sheath around the nebuliser flow on the sensitivities (open bars) and detection limits (dashed bars). The nebuliser flow rate was 0.45 L.min^{-1}.

Table 5. Background count rates (counts s^{-1} ± % relative standard deviation) observed for 0.1 M HNO_3 with a sheath of nitrogen or hydrogen

Analyte ion	N_2 trace		H_2 trace		H_2 0.05 $L.min^{-1}$	
	counts	RSD	counts	RSD	counts	RSD
^{51}V	49.7	16	29.5	28	126	23
^{52}Cr	927	10	625	3	648	2
^{55}Mn	3760	13	483	3	728	7
^{56}Fe	90100	6	123200	3	92300	2
^{57}Fe	461	3	1100	4	1110	4
^{60}Ni	55.5	14	2200	50	30.2	21
^{63}Cu	309	11	387	6	134	3
^{66}Zn	363	11	730	17	111	10
^{76}Se	3700	7	2810	1	2360	3
^{114}Cd	108	8	700	35	46	42
^{208}Pb	190	5	134	10	17.0	14

Table 6. Signal to background ratios observed for 100 $ng.mL^{-1}$ of each element in 0.1 M HNO_3, with and without a sheath of gas

Analyte ion	No sheath	Ar 0.5 $L.min^{-1}$	Ar 1.0 $L.min^{-1}$	N_2 trace	H trace	H_2 0.05 $L.min^{-1}$
^{51}V	9400	5200	31	2100	4400	300
^{52}Cr	220	83	0.2	9	180	49
^{55}Mn	300	190	0.8	30	300	54
^{56}Fe	1.8	0.8	N.D.	1.0	1.0	0.4
^{67}Fe	2.8	1.5	0.1	5.1	2.8	0.8
^{60}Ni	120	150	17	340	11	260
^{63}Cu	120	100	7	140	130	120
^{66}Zn	29	30	9	37	17	44
^{76}Se	7.5	0.3	0.006	0.4	0.5	2.0
^{114}Cd	770	1400	5	240	43	200
^{208}Pb	260	330	4	140	240	480

N.D. = not determined.

The increase in background comes from the greater proportion of argon (as well as its impurities)[11] being sampled through the interface which leads to higher $^{36}Ar^{16}O^+$, $^{40}Ar^{12}C^+$, $^{38}Ar^{38}Ar^+$ and $^{40}Ar^{36}Ar^+$ count rates.

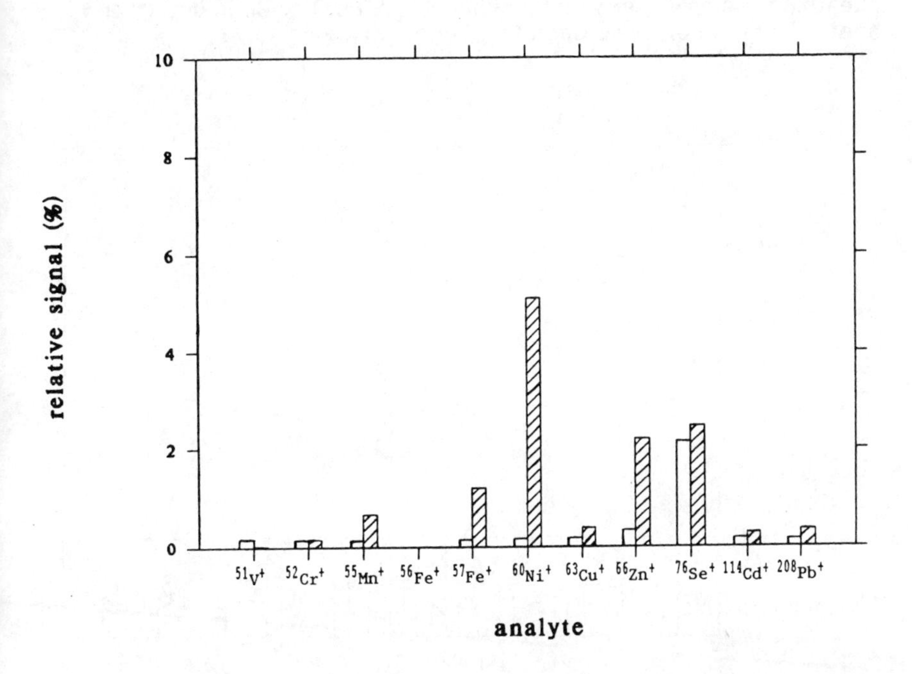

Figure 3 Effect of the addition of 1.0 L.min^{-1} Ar as a sheath around the nebuliser flow on the sensitivities (open bars) and detection limits (dashed bars). The nebuliser flow rate was 0.08 L.min^{-1}.

Effect of Nitrogen

The relative sensitivities and detection limits obtained with a sheath of nitrogen are reported in Table 3 and shown graphically in Figure 4. The nitrogen was introduced in trace amount (*i.e.* less than 0.05 L.min^{-1}, the smallest reading of the flow meter) because the plasma was found to extinguish upon introduction of more nitrogen. The optimum nebulizer gas flow rate under these conditions was 0.6 L.min^{-1}. As with a sheath of argon, all the sensitivities were degraded upon the introduction of nitrogen, with the exception of

$^{66}Zn^+$ which remained essentially the same. As with 0.5 $L.min^{-1}$ of argon, most of the detection limits were degraded. However, a significant enhancement was observed for $^{60}Ni^+$ which showed the same improvement as with 0.5 $L.min^{-1}$ of argon compared with the usual operating conditions free of sheathing gas. The detection limits for $^{57}Fe^+$, $^{66}Zn^+$ and $^{208}Pb^+$ remained essentially the same as without sheathing, being either slightly enhanced ($^{57}Fe^+$) or degraded ($^{66}Zn^+$ and $^{208}Pb^+$) under these operating conditions.

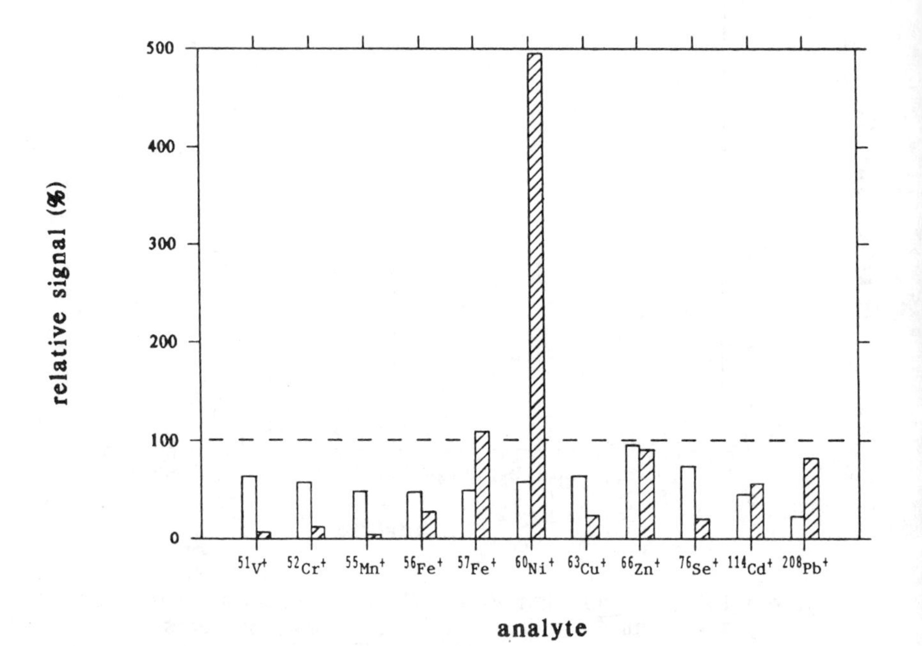

Figure 4 Effect of the addition of a trace of nitrogen as a sheath around the nebuliser flow on the sensitivities (open bars) and detection limits (dashed bars). The nebuliser flow rate was 0.06 $L.min^{-1}$.

A small addition of nitrogen appears to have a significant quenching effect on $^{40}Ar^{16}OH^+$ but much less on $^{40}Ar^{16}O^+$ since the background under $^{56}Fe^+$ is only slightly reduced (Tables 4 and 5) compared with no sheathing gas. This probably explains why an improvement was seen for $^{57}Fe^+$ but not $^{56}Fe^+$. Tables 4 and 5 show a significant increase in background count rates for $^{51}V^+$, $^{52}Cr^+$, $^{55}Mn^+$, $^{76}Se^+$ and $^{114}Cd^+$, as well as a general degradation of RSD (except for $^{60}Ni^+$, $^{114}Cd^+$ and $^{208}Pb^+$). The increase in background at amu 51 and 55 can probably be attributed to increased $^{36}Ar^{15}N^+$ and $^{40}Ar^{15}N^+$ caused by the addition of nitrogen while that at amu 52 may be caused by higher $^{36}Ar^{16}O^+$ and/or $^{40}Ar^{12}C^+$ originating from oxygen and/or carbon impurities in nitrogen. The higher background than that observed without a sheath at amu 76 and 114 cannot be explained by isobaric interferences from polyatomic species arising from nitrogen. Table 6 shows more degraded S/B for all the isotopes except $^{57}Fe^+$, $^{60}Ni^+$, $^{63}Cu^+$ and $^{66}Zn^+$ when compared with the results from both no sheath or 0.5 L.min^{-1} Ar. However, a sheath of nitrogen enhanced the S/B of $^{57}Fe^+$, $^{60}Ni^+$, $^{63}Cu^+$ and $^{66}Zn^+$, the biggest improvement being seen for $^{60}Ni^+$ (three-fold).

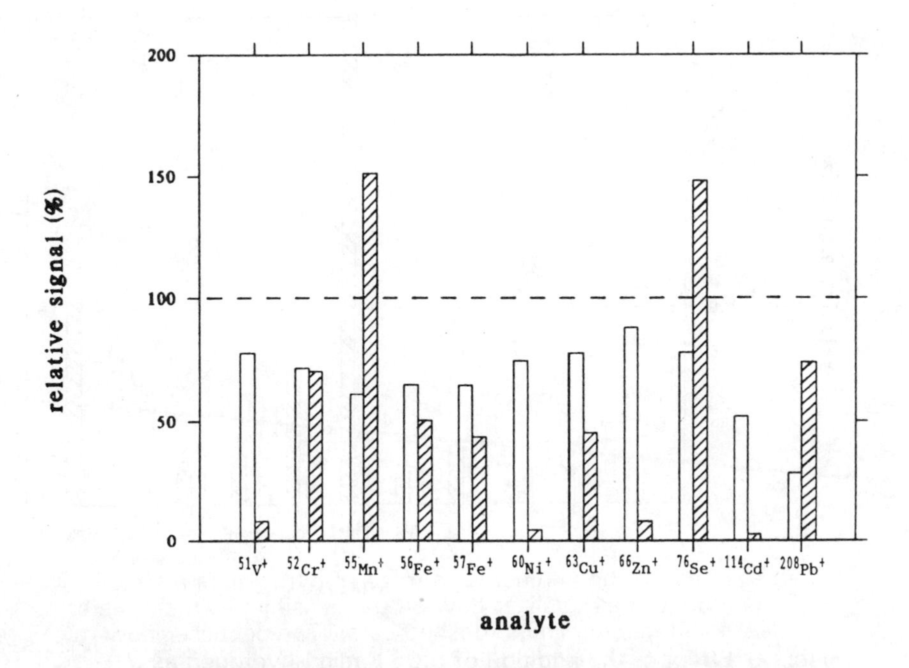

Figure 5 Effect of the addition of a trace of hydrogen as a sheath around the nebuliser flow on the sensitivities (open Bars) and detection limits (dashed bars). The nebuliser flow rate was 0.75 L.min^{-1}

The observations made above for a sheath of nitrogen are completely different from those reported for mixed gas plasmas produced by the addition of nitrogen to either the nebulizer gas[9] or the plasma gas[15] (*i.e.* the bulk of the plasma). In the former case[9], the addition of nitrogen to the nebulizer gas led to a decrease in background count rate of the argon dimers and improved S/B for $^{76}Se^+$ and $^{78}Se^+$ as a smaller proportion of argon was being sampled from the injection channel of the plasma. In the present study, nitrogen is added as a sheath around this injection channel which is the main region sampled through the interface. A much smaller injection channel and/or reduced nebulizer flow rate would therefore be required in order to increase the amount of sheathing gas being sampled, thereby reducing the proportion of argon.

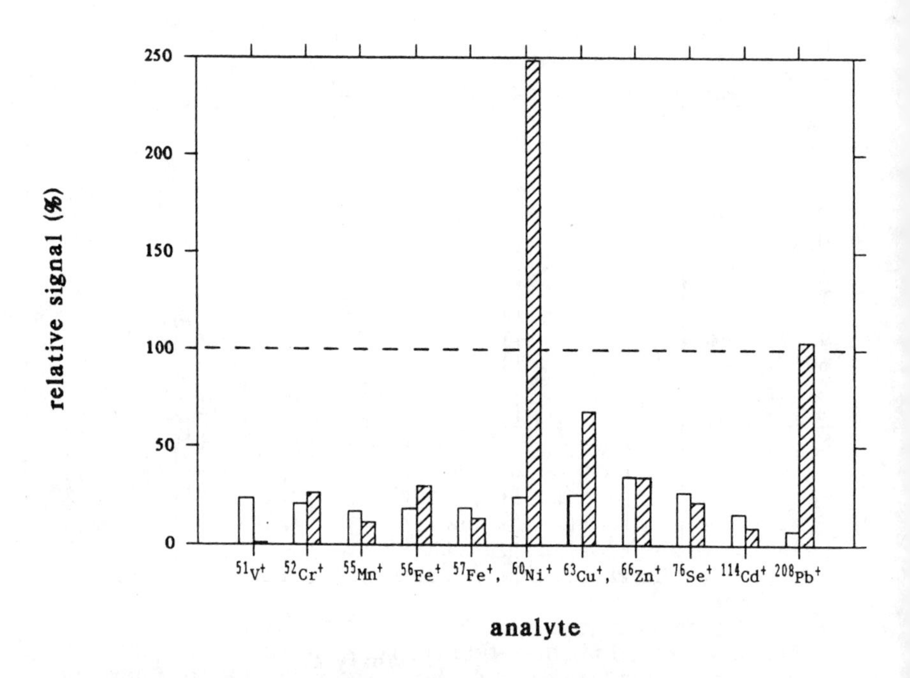

Figure 6 Effect of the addition of 0.05 L.min^{-1} hydrogen as a sheath around the nebuliser flow on the sensitivities (open Bars) and detection limits (dashed bars). The nebuliser flow rate was 0.65 L.min^{-1}.

One of the main changes caused by the addition of a sheathing gas is an increase in the residence time of the analyte inside the plasma. This is in contrast with the addition of nitrogen to the plasma gas where a decrease of the physical size of the plasma discharge was observed[15]. This decrease in size moved the initial radiation zone away from the sampler, and a higher nebulizer flow rate, as well as a smaller sampling depth, were then required to obtain optimum sensitivity[15]. Using 1.5 kW, 5% nitrogen in the plasma gas, 1.25 $L.min^{-1}$ nebulizer flow rate and sampling 8 mm above the load coil, improvement in sensitivity, by a factor of up to 4, was observed for $^{64}Zn^+$, $^{73}Ge^+$, $^{75}As^+$, $^{64}Zn^+$, $^{88}Sr^+$, $^{89}Y^+$, $^{90}Zr^+$ and $^{99}Ru^+$ [15].

Effect of hydrogen

When hydrogen was introduced as a sheath in trace amounts, the optimum nebulizer gas flow rate was 0.75 $L.min^{-1}$, slightly higher than for a sheath of nitrogen. It had to be decreased to 0.65 $L.min^{-1}$ when the flow rate of hydrogen was increased to 0.05 $L.min^{-1}$ (the highest that could be used without extinguishing the plasma). The results obtained at these two sheathing flow rates are summarized in Table 3 and shown graphically in Figures 5 and 6, respectively. All sensitivities were lower with a trace of hydrogen than without sheathing, but were slightly better than with a sheath of nitrogen. They however decreased substantially upon addition of more hydrogen even if the sampling depth was readjusted by decreasing the nebulizer flow rate. This is in contrast with results obtained in ICP-AES[7] where the maximum intensity for ionic lines was enhanced and shifted to a lower observation height.

Compared to usual operating conditions without sheathing, adding a trace of hydrogen improved detection limits only for $^{55}Mn^+$ and $^{76}Se^+$ (Tables 2 and 3) due to significantly improved background stability for these elements. Tables 4 and 5 show that improvement in RSDs were only seen for those two elements whereas lower background count rates were observed for $^{52}Cr^+$, $^{55}Mn^+$, $^{57}Fe^+$, $^{63}Cu^+$, and $^{208}Pb^+$. Detection limits were degraded by more than 90% for $^{51}V^+$, $^{60}Ni^+$, $^{66}Zn^+$, and $^{114}Cd^+$, which corresponded to an increase in background as well as a dramatic decrease of their S/B. This is the only case in this work where the detection limit for $^{60}Ni^+$ was degraded upon the introduction of a sheathing gas.

However, when the hydrogen flow rate was increased to 0.05 $L.min^{-1}$, the background count rate (Tables 4 and 5) either remained similar to that without a sheath or decreased substantially, for all the analytes except $^{51}V^+$. This observation, combined with lower RSDs (Table 4 and 5) and improved S/B (Table 6) explains the improved detection limit for $^{60}Ni^+$, being 2.5 times better than without sheathing (Table 2). As can be seen from comparing Figure 5 and Figure 6, as well as in Table 3, the detection limit for $^{60}Ni^+$ improved by more than a factor of 200 upon an increase in the flow rate of hydrogen. This also resulted in improved detection limits for $^{51}V^+$, $^{56}Fe^+$, $^{57}Fe^+$, $^{63}Cu^+$, $^{66}Zn^+$, $^{114}Cd^+$ and $^{208}Pb^+$, but degraded detection limits for both $^{55}Mn^+$ and $^{76}Se^+$. Compared with usual operating conditions, a small improvement was observed for $^{208}Pb^+$.

Implications for mass discrimination

Since varying the nebulizer flow rate and the gas composition changes the expansion pressure, the effect of sheathing gases on the instrumental mass response was evaluated. As can be seen from Table 7, all of the sheathing gases lowered the degree of mass discrimination by flattening the response observed over the whole mass range. The smallest change was obtained using argon as the sheathing gas whereas the most uniform response was observed with hydrogen at a flow rate of 0.05 $L.min^{-1}$. A small degree of mass discrimination is however advantageous in isotope ratio and isotope dilution works, since it could eliminate the need to correct for changes in sensitivities between subsequent amu values.

The mass response with or without a sheathing gas is different than that reported previously[11] where the maximum response was obtained around 100 amu with a slow decrease towards higher masses and a much more rapid drop on the lower masses side. The reason for the difference is unknown at the moment of writing and will be the subject of investigation. Nonetheless, the above observations about a flattening of the mass response still hold since all the measurements were made with the ion lenses set such that the response without a sheathing gas was the same from day to day.

Table 7. Molar sensitivity corrected to 100% abundance and expressed (in %) relative to that of Mn for each condition with and without a sheathing gas

Analyte ion	No sheath	Ar 0.5 $L.min^{-1}$	N_2 trace	H_2 trace	H_2 0.05 $L.min^{-1}$
^{51}V	64.7	65.7	86.4	82.7	90.3
^{52}Cr	73.4	73.1	88.2	86.5	90.8
^{55}Mn	100	100	100	100	100
^{56}Fe	92.6	95.1	92.0	98.5	100
^{57}Fe	91.6	93.9	94.4	97.2	101
^{60}Ni	55.3	58.4	67.6	67.5	80.2
^{63}Cu	47.1	50.8	62.9	59.9	70.5
^{66}Zn	25.6	29.0	51.0	36.8	52.5
^{76}Se	12.3	14.0	19.0	15.7	19.4
^{114}Cd	172	167	165	147	164
^{208}Pb	352	290	173	162	147

4 CONCLUSIONS

In general, the introduction of a sheathing gas in ICP-MS decreased the sensitivity, but improved the stability to the point where equal or better detection limits could be obtained for several analytes ($^{55}Mn^+$, $^{57}Fe^+$, $^{60}Ni^+$, $^{66}Zn^+$, $^{76}Se^+$, $^{114}Cd^+$ and $^{208}Pb^+$). The detection limits of $^{55}Mn^+$ and $^{76}Se^+$ were slightly improved (by a factor of 1.5) with a trace of hydrogen (as opposed to no sheathing). The addition of a trace of nitrogen led to similar detection limits for $^{57}Fe^+$ and $^{66}Zn^+$ but to a five-fold improvement for $^{60}Ni^+$. The same improvement in the detection limit of $^{60}Ni^+$ was obtained with 0.5 $L.min^{-1}$ Ar, as well as factors of 3.7 and 1.9 improvement for $^{114}Cd^+$ and $^{208}Pb^+$, respectively. In all cases, the nebulizer flow rate had to be decreased in order to observe maximum sensitivity, for a given sheathing gas flow rate. The improvement in stability is even more impressive if the poor flow meter used to regulate the sheathing gas is taken into account. Even bigger improvements would potentially be obtained if a mass flow controller was used for the sheathing gas.

The mass discrimination was decreased by the addition of a sheathing gas, which indicates that the change in expansion pressure induced by the combination of a lower nebulizer flow rate and a different gas composition, is changing the space charge downstream from the skimmer, and results in a more focused ion beam. Gillson and coworkers[16] showed that the transmission characteristics of the ion optics are strongly influenced by the space charge of the ion beam. Any change in the flux and/or composition of the ion beam will affect the space charge downstream from the skimmer which will result in modified ion trajectories and signals[16]. The data in Table 7 suggests that introducing a sheath of gas around the nebulizer flow has a "focusing" effect on the ions extracted. However, why a particular gas has a more pronounced effect on some analytes than others is a question which will be addressed in future work.

Finally, the fact that ions are still detected at a very low nebulizer flow rate shows potential for using a sheathing device as a means of coupling a GC to ICP-MS. Since a typical effluent flow rate in GC is about 50 $mL.min^{-1}$, a make-up gas is required to puncture the plasma. However, if a tee is used for this purpose, the effluent is mixed (and the analyte thereby diluted) with the make-up gas. The results reported in this work seem to indicate that using a sheathing device to add the make-up gas would limit the dilution of the analyte, the make-up gas remaining more or less a sheath around the effluent from the GC column. Work is in progress to verify this observation.

ACKNOWLEDGEMENTS

The principal author gratefully acknowledges the financial support of the Natural Sciences and Engineering Research Council of Canada (Grant # OGP0039487) and of the Advisory Research Committee of Queen's University.

REFERENCES.

1. G.M. Hieftje and G.H. Vickers, *Anal. Chim. Acta*, 1989, **216**, 1.
2. M.A. Vaughan and G. Horlick, *Appl. Spectrosc.*, 1986, **40**, 434.
3. S.H. Tan and G. Horlick, *Appl. Spectrosc.*, 1986, **40**, 445.
4. M.E. Ketterer, J.J. Reschl and M.J. Peters, *Anal. Chem.*, 1989, **61**, 2031.
5. J.T. Rowan and R.S. Houk, *Appl. Spectrosc.*, 1989, **43**, 973.
6. D.J. Douglas, *Can. J. Spectrosc.*, 1989, **34**, 38.
7. M. Murillo and J.M. Mermet, *Spectrochim. Acta*, 1989, **44B**, 359.
8. G. Horlick, Paper #PL5, 1988 Winter Conference on Plasma Spectrochemistry, San Diego, California, January 3-9.
9. E.H. Evans and L. Ebdon, *J. Anal. At. Spectrom.*, 1989, **4**, 299.
10. A. Goldwasser and J.M. Mermet, *Spectrochim. Acta*, 1986, **41B**, 725.
11. D. Beauchemin, J.W. McLaren and S.S. Berman, *Spectrochim. Acta*, 1987, **42B**, 467.
12. S.R. Koirtyohann, J.S. Jones, C.P. Jester and D.A. Yates, *Spectrochim. Acta*, 1981, **36B**, 49.
13. F.E. Lichte, A.L. Meier and J.G. Crock, *Anal. Chem.*, 1987, **59**, 1150.
14. M. Murillo and J.M. Mermet, *Spectrochim. Acta*, 1987, **42B**, 1151.
15. J.W.H. Lam, Ph.D. Thesis, Edmonton, Alberta, Canada, 1988, pp. 28-71.
16. G.R. Gillson, D.J. Douglas, J.E. Fulford, K.W. Halligan and S.D. Tanner, *Anal. Chem.*, 1988, **60**, 1472.

ANALYTICAL PERFORMANCE OF ANALOGUE DETECTION IN ICP-MS

R.C. Hutton, A. Walsh and R.N. Gosland

VG Elemental Limited
Ion Path
Road Three
Winsford
Cheshire CW7 3BX UK

1 INTRODUCTION

The detection systems currently in use in commercial inductively coupled plasma mass spectrometry (ICP-MS) instruments usually employ pulse counting electron multipliers. These devices operate in a high gain mode and produce high signal outputs from small ion fluxes. This mode of detection is commonly used in many measurement systems for small signal levels. In ICP-MS, however, the sensitivity of the instruments has increased to such an extent that the use of pulse counting detection has become less necessary in obtaining low detection limits.

One disadvantage of such a sensitive detection system is that, at moderately high ion fluxes, the device will produce a signal with a non-linear dependance on input flux, this being a direct consequence of gain suppression in the detector. In practical terms, this limits the upper working range in conventional pulse counting ICP-MS to signals of approximately 106 counts per second which often corresponds to concentrations in solution of under 1 μg.mL^{-1}. In order to overcome this inconvenience, commercial instruments are equipped with multiplier protection logic to guard against high ion fluxes and a variety of measures have been proposed to extend this analytical dynamic range further. These include:

1. attenuation of the incident ion beam using:
   a. the ion optics
   b. the quadrupole dc offset
2. in-line dilution
3. use of low gain (analogue) detection

This paper describes some of the performance characteristics of the last arrangement and suggests a revision of the philosophy of

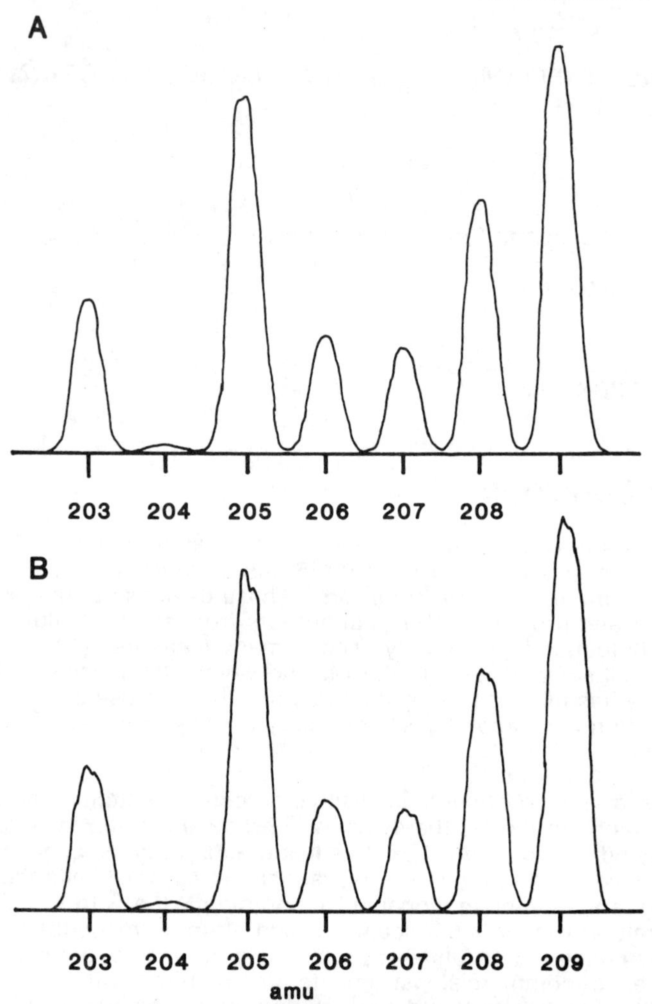

Figure 1 Analogue (A) and pulse counting (B) spectra for (A) 100 ng.mL^{-1} and (B) 100 μg.mL^{-1} of Tl, Pb and Bi. Aquisition time for each spectra is 15 secs.

detection in ICP-MS.

2 DISCUSSION

Characteristics of analogue detection

The ions transmitted by the quadrupole are detected by an electron multiplier assembly. This device is usually a continuous dynode multiplier consisting of a glass tube with a resistive coating of typically 108 ohms. A voltage of approximately -3kV is applied to the multiplier to cause ions to enter the funnel. The gain of such a device is typically 107 *i.e.* for each incident ion a pulse of typically 107 electrons is produced at the output end. The lifetime of this multiplier will be dependent on the total accumulated charge which is proportional to accumulated counts[1]. In order to operate at higher ion fluxes, without therefore damaging the device, a multiplier operating in a low gain analogue mode is employed in the VG PlasmaQuad ICP-MS instrument. Applying a voltage of around -1.5 kV produces a gain of 103 to 104 and in this mode, the multiplier is not saturated, so the overall gain, and the resulting output pulse size, depends upon the applied voltage. The electrons are collected in the conventional manner and measured as mean current using a linear amplifier, the response time of which is sufficiently fast so as to retain the rapid data acquisition characteristics of the instrument. Since the lifetime of the detector depends on the total accumulated charge, analogue operation will obviously be beneficial in obtaining a longer detector lifetime.

A further advantage of the analogue mode, when measuring high concentrations in samples, is that unlike any beam attenuation systems which operate on the quadrupole, the mass spectrometer performance is not compromised; whereas in systems where a quadrupole parameter is changed, peak shapes, resolution and mass discrimination may be adversely affected. Figure 1 illustrates a spectrum in both analogue and pulse counting modes and the retention of peak shape quality is apparent.

Linearity and precision with analogue detection

The precision of measurement in analogue mode is, as in pulse counting, a function of integration time. Typically, precisions of <1% RSD can be achieved with moderate total integration times of approximately 0.5 second per isotope. This is shown in Figure 2 where it can also be deduced that further increases in this time produces no further significant gain in precision.

Using a combination of analogue and pulse counting detection, therefore a linear dynamic range spanning eight orders of magnitude can be achieved[2]. The instrument software is designed to perform a

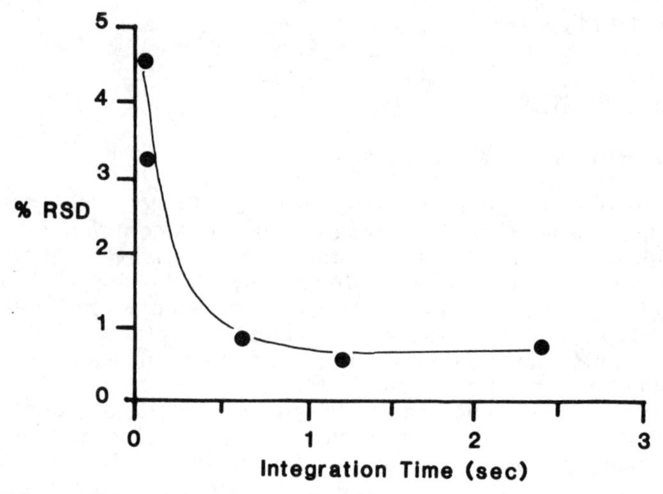

Figure 2 Precision in analogue mode as a function of integration time for 20 $ng.mL^{-1}$ La

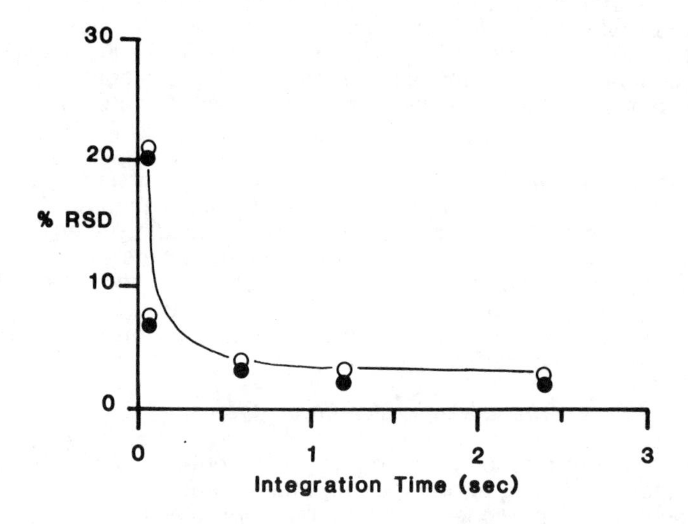

Figure 3 Analogue mode precision close to the detection limit for 10 $ng.mL^{-1}$ La (open symbols) and 20 $ng.mL^{-1}$ La (filled symbols).

cross calibration between the two modes and for the user of the instrument, the detection can therefore be decided intelligently by the computer depending on the count rate of the analyte isotope.

Analytical performance of dual mode detection

Automatic dual mode detection will allow the analysis of a sample with a wide concentration range of elements and produce good accuracy and precision with minimal operator interaction. Table 1 illustrates the analysis of an urban dust standard reference material (SRM), produced by the National Institute for Standards and Technology (NIST) which contains analytes over a wide range of concentrations. A description of the analysis of majors, minors and traces in geological samples can be found in a VG Technical Information sheet[3].

Table 1 Analysis of SRM NIST 1648 urban particulate

Element	Mode	Certified (wt%)	ICP-MS (wt%)
	Analogue		
Al		3.42	3.80
K		1.05	0.90
Fe		3.91	3.96
Zn		0.48	0.46
Pb		0.655	0.70
	Pulse Counting	($\mu g.g^{-1}$)	($\mu g.g^{-1}$)
Ag		6	5
Ba		740	800
Cd		75	75
Ce		55	50
Co		18	15
Cu		609	605
La		42	50
Rb		52	57
Th		7.4	7.5
U		5.5	4

Analogue detection

The detection limits achievable in analogue mode alone are similar to those documented for sequential ICP-optical emission spectrometry (ICP-OES) as shown in Table 2. There are some

Table 2 Comparative detection limits (ng.mL^{-1}) for ICP-AES and ICP-MS

Element	ICP-AES	VG PlasmaQuad ICP-MS Pulse Counting	Analogue
Li	0.9	0.02	1.0
Mg	0.08	0.02	1.5
Mn	0.4	0.02	0.9
Ba	0.1	0.02	1.2
Pb	20	0.01	1.0
Al	4	0.02	1.1
Cu	0.9	0.02	1.3
Cr	2	0.05	1.0
La	ND	0.002	0.8

ND = not determined

elements also which are more usefully determined in analogue mode, these being elements which suffer from spectral overlap such as $^{28}Si/N_2$, $^{31}P/NOH$, $^{32}S/O_2$[4] *etc*. Using pulse counting detection in these cases, the overlap can be a relatively large proportion of the calibration. Hence in pulse counting with only a maximum of approximately 106 counts per second (cps) then an overlap, such as N^{2+} or O^{2+} which can be 104 or 105 cps will seriously limit analytical accuracy and precision, it being the dominant source of error in such a measurement.

With analogue detection, the size of the overlap relative to the

Table 3 ICP-MS detection limits for some difficult elements

Element	Pulse Counting	Analogue
Si	10	25
P	5	10
S	600	350
Cl	70	100
K	5	1.5
Ca	3	10
Fe	0.1	1

3σ values (ng.mL^{-1})

maximum calibration is less, hence the overlap will only dominate the precision at lower concentrations. It is particularly fortunate that these elements are generally required to be determined at higher levels. Detection limits for some of these elements in analogue mode are given in Table 3.

It can thus be concluded that, in certain situations, analogue detection alone is the best mode for an analysis. This would have the advantage of:

1. being a single measurement cycle
2. eliminating the need for complex multiplier protection circuitry.
3. being similar in detection to ICP-OES but with the spectral simplicity of ICP-MS

Table 4 Environmental analysis: analogue detection limits for EPA method 6020[5]

Element	EPA 6020 requirements	ICP-MS analogue mode
Be	5	0.2
Cr	10	1
Mn	15	1
Ni	40	1
Cu	25	0.5
Zn	20	1.5
As	10	1
Se	5	5
Ag	10	1
Cd	5	1
Hg	0.2	1.5
Ti	10	1
Pb	5	1
Na	5000	20
Mg	5000	1
Al	200	1
K	5000	10
Ca	5000	15
V	50	1
Fe	100	5
Co	50	0.2
Sb	60	1
Ba	200	1.5

Results in $ng.mL^{-1}$

The analogue only measurement is of course not applicable to all sample types. However, it can be effectively employed especially in certain environmental analytical problems where the ultimate in detection is not required. Hence, where analytical requirements are the range of 10 $ng.mL^{-1}$ and above, analogue detection can in many instances perform as well as pulse counting.

Certainly even at levels close to the detection limit, the precision achievable is of the order of a few percent as seen in Figure 3.

3 ENVIRONMENTAL ANALYSIS

Many environmental samples such as waters, sludges, and sediments require to be monitored for a wide range of metallic impurities. In many cases the composition is unpredictable and hence, any analytical technique needs to be sufficiently robust to reliably measure such samples. The US Environmental Protection Agency (US EPA) sets out within their contract lab program (CLP) a range of such elements which must be determined. An analytical protocol (6020) is being

Table 5 Analogue detection for environmental analysis: results for SRM NIST 1643B ($ng.mL^{-1}$)

Element	Certified	ICP-MS Analogue Mode
Be	19	20.2
B	(94)	85
Cr	18.6	18.5
Mn	28	27.9
Fe	99	130
Co	26	24.5
Cu	22	20.2
Zn	66	64
As	(49)	55
Se	9.7	<10
Sr	227	219
Mo	85	86
Ag	9.8	10
Cd	20	18.5
Ba	44	40
Pb	23.7	20
Bi	11	9

() = recommended value only

validated by EPA to allow ICP-MS to be used for such analyses. The elements and detection limits are given in Table 4. In this case, the detection requirements are generally not severe and analogue detection is capable of achieving most of those required levels.

Quantitative analyses of such environmental samples using only analogue detection produces (in general) similar results to certified values as can be seen from Table 5.

4 CONCLUSIONS

The use of analogue detection both in combination with pulse counting and alone offers the analyst considerable flexibility for ICP-MS analysis. The combination of pulse counting and analogue, yields a dynamic range of up to eight orders allowing majors, minors and traces to be determined with good precision[2]. The intelligent software incorporated into the system provides automatic selection of detection mode based on peak intensities. Hence the need for operator interaction on unknown samples is minimised.

Analogue detection alone provides detection limits easily comparable to ICP-OES with the added convenience of isotopic spectral simplicity. Hence for ICP-MS analyses where detection to sub ppb levels are not required or where an inconvenient spectral overlap exists, the analogue mode is a more flexible detection alternative, combining as it does, sensitivity and dynamic range whilst minimising the need for inconvenient multiplier protection.

In conclusion, for analysis at below ppb levels, pulse counting may be considered to be the best detection mode in ICP-MS, however, at levels from 10 $ng.mL^{-1}$ to sub %, the use of analogue detection can, in many instances, be a much more attractive option.

REFERENCES

1. R.H. Prince and J.A. Cross, *Rev. Sci. Instrum.* 1971, **42**, 66.
2. R.C. Hutton, A.N. Eaton and R.M. Gosland., *Appl. Spectrosc.* 1990, **44**, 238.
3. VG Technical Information PQ 802 (1989)
4. S.H. Tan and G. Horlick, *Appl. Spectrosc.*, 1986, **40**, 445.
5. US Environmental Protection Agency, contract laboratory program, statement of work for inorganic analysis No. 788

THE DETERMINATION OF TITANIUM, COPPER, AND ZINC IN GEOLOGICAL MATERIALS BY INDUCTIVELY COUPLED PLASMA MASS SPECTROMETRY WITH MULTIVARIATE CALIBRATION

Michael E. Ketterer[1], John J. Reschl[1] and Michael J. Peters[2]

[1]U.S. Environmental Protection Agency,
National Enforcement Investigations Center,
Box 25227, Building 53, Denver Federal Center,
Denver, Colorado 80225 USA

[2]ICF Technology Inc.,
165 S. Union Boulevard Suite 802,
Lakewood, Colorado 80228 USA

1 INTRODUCTION

Inductively coupled plasma mass spectrometry (ICP-MS) has become an established method of multi-element trace analysis[1]. This technique offers favourable characteristics such as low detection limits for a large number of elements, rapid determination of elemental concentrations, and the ability to provide rapid, accurate isotopic information with minimal sample preparation. Based upon these characteristics, ICP-MS's performance compares favourably to that of flame and furnace atomic absorption spectroscopy, inductively coupled plasma atomic emission spectroscopy, and other types of mass spectrometry for a large variety of applications.

The existence of isobaric polyatomic ion interferences was established early in the development of ICP-MS[2-4], and represents a significant shortcoming of the technique. These interferences are due to the formation of species such as metal oxides, hydroxides, dioxides, and chlorides which appear at the same nominal mass as an analyte of interest. Polyatomic species may arise from the sample itself, the solvent, reagents added to the sample, or from the gases used to form the plasma. Other species such as doubly charged and dimeric ions are known, and these can cause similar problems. In addition to polyatomic isobaric interferences, elemental isobaric interferences (*e.g.* Nd and Sm at 148 amu) exist. Nearly all of these isobaric interference problems are unresolvable using the quadrupole mass analyzer. As a result, researchers have attempted to circumvent these interferants through mechanisms such as adjustment of plasma conditions[5], aerosol desolvation[6], argon-nitrogen plasmas[7], and chemical separations of interferants[8]. A technique based upon commonly used "elemental equations" has been proposed where the degree of MO^+/M^+ is estimated through the use of an internal standard for oxide formation[9]; the signal due to MO^+ is then subtracted from that of M'^+, the analyte. Each of these strategies offers some promise for alleviating difficulties due to polyatomic ion interferences; however, none is entirely satisfactory and/or successful in all conceivable cases.

Recently, multivariate calibration methods such as multiple linear regression, the generalized standard addition method, principal components regression, and partial least-squares regression have been widely used in chemical analysis[10-13]. These techniques all utilize information from multiple, partially selective sensors in order to determine analyte(s) in the presence of interferant(s). The techniques mentioned above differ principally in the manner in which one manipulates the calibration data and performs unknown predictions.

Multiple linear regression and principal components regression have been applied to ICP-MS for the determination of Cd in the presence of Mo, and the determination of Cd, In, and Sn in the presence of Zr, Mo, and Ru[14]. It was found that both MLR and PCR are viable approaches for these types of determinations, and that they may be widely applicable to interference situations in ICP-MS. The purpose of the present study is to provide a detailed description of how one applies MLR to a real-sample problem, namely, the determination of Cu and Zn in soil and rock materials which contain high levels of Ti. The use of multiple linear regression with external calibration, and standard addition will be demonstrated.

2 IMPLEMENTATION OF MULTIVARIATE CALIBRATION

A general mechanism for implementing multivariate calibration in ICP-MS consists of the following steps:

(1) survey of the problem;
(2) choice of calibration methodology;
(3) design of the calibration set;
(4) analysis of the calibration set;
(5) analysis and prediction of unknowns.

In step (1), the analyst's objective is to define which species cause signals at the sensors (masses) available for analyte determination. One must consider sample-, solvent-, reagent- and gas-related polyatomic species. Partial factorial experiments such as Plackett-Burman designs[15] offer an effective tool for determining whether or not a species contributes to the signal at a given mass. In such experiments, one does not execute all possible variable combinations, but a chosen subset of them. To implement such an experiment, one systematically adds potential interferants to the sample, measures the resulting signals at masses of interest, and uses statistical tests to discern whether or not a given addition yields a significant change in the response. Through proper selection of the partial factorial design, one can examine a large number of potential interferants in a minimal number of trials. An example of this process is seen in subsequent sections, where the effects of 10 potential sources of signal at masses 64 and 66 are investigated using 16 trial additions.

The choice of calibration methodology (Step 2) requires several considerations on the part of the analyst. In general, multivariate calibration need not be used if the problem is solvable through other means; for example, it is possible to apply multivariate calibration to the determination of Nd and Sm using only masses 148 and 150. However, it is much simpler to determine Nd and Sm using fully selective sensors at 146 and 147 amu, respectively. Furthermore, many potential interference situations are simply not of concern in real samples; the influences of osmium, iridium, and platinum oxides at masses 206-208 are unlikely to be significant in the determination of lead in urban soil samples.

Although many multivariate calibration methods have been developed, three methods are commonly applied to similar problems in other areas of analytical chemistry: multiple linear regression (MLR), principal components regression (PCR) and partial least squares regression (PLSR). Beebe and Kowalski[13] have discussed the differences between these approaches at length. MLR involves developing a calibration model which directly associates the responses at the sensors with concentrations of analytes and interferants in the calibration set. An example of a set of MLR calibration equations for a problem with two sensors (m1 and m2) and two analyte concentrations (c_j and c_k) is shown below:

$$R_{m1} = a_{m1}c_j + b_{m1}c_k + i_{m1} \quad (1)$$

$$R_{m2} = a_{m2}c_j + b_{m2}c_k + i_{m2} \quad (2)$$

In these expressions, R_{m1} and R_{m2} refer to the responses at sensors m1 and m2; a and b are the regression coefficients (slope terms) at sensors m1 and m2 for analytes j and k; and i_{m1} and i_{m2} are the intercepts at m1 and m2.

In PCR, one first applies principal components analysis to the calibration set responses, producing a group of new variables known as principal components. These principal components are a more succinct expression of the variance in the calibration set; the principal components analysis process combines correlated variables and separates uncorrelated variables. The values of these new variables, referred to as principal component scores, are then modelled in terms of analyte/interferant concentrations in the calibration set using simple or multiple linear regression. PLSR basically involves a principal components analysis of both the response and concentration matrices of the calibration set, followed by simple or multiple regression to associate the concentration and response components. Generally speaking, MLR is widely applicable to most potential interference problems in ICP-MS because of its simplicity. MLR is probably the approach preferred in most situations, but it is difficult to execute for highly complex systems, and yields poor results where collinearity exists in the responses. The latter problem occurs in situations where the relative responses of two analytes at two sensors are similar. For example, collinearity would be a problem in the determination of Cd in the presence of Mo if one used only 110 and 112 amu; the $^{112}Cd/^{110}Cd$ ratio of 1.94 is close to the $^{96}Mo/^{94}Mo$ ratio (1.84). In such circumstances, the measurement system is unable to identify the source of a signal from the available sensors. On the other hand, the determination of Cd in the presence of Zr using only 110 and 112 amu would be readily accomplished by MLR since the $^{96}Zr/^{94}Zr$ ratio of 0.16 is vastly different from 1.94.

Principal components regression is better suited for complex situations of more than three or four signal sources, and where collinearity exists; PLSR has the same advantages as PCR but is also applicable to under-determined systems (fewer sensors than analytes/interferants)[13]. PLSR also provides a measure of the applicability of the calibration set to an unknown sample, which is very useful in warning where unanticipated sources of signal in a sample have not been considered. Nevertheless, MLR is expected to be very useful for most isobaric interference situations in ICP-MS.

An additional matter to be decided in Step 2 is whether one wishes to use an external calibration or standard addition scheme. Either one may be used in combination with MLR, PCR, or PLSR. Standard addition incorporated into a multivariate calibration problem is referred to as the "generalized standard addition method" or GSAM. In GSAM, the calibration set is run in the sample itself, and the results are analyzed by MLR, PCR, or PLSR. The GSAM-MLR combination would be executed by using the concentrations of unknowns added to the sample instead of concentrations in the external standards. GSAM is practical where a limited number of samples are to be analyzed, or in situations where the matrix suppression of analyte signals is severe and internal standards do not emulate the analytes' responses. Generally, external calibration standards are preferred on the basis of sample throughput.

Having surveyed the problem and chosen a calibration methodology, a design is constructed for the calibration set. The subject of experimental design is relevant to this process, as one is faced with the problem of systematically varying several parameters so that their effects may be resolved in as few trials as possible. Generally, one can employ Plackett-Burman designs, use two or three concentration levels for each analyte, perform some replication, and execute between n+1 and 5n calibration standards (n= the number of analytes and interferants) in a complete calibration set. It is of importance to construct a design which permits the identification of the source of signals. In a problem with two sources of signal, for example, a calibration design which consists of points at (0,0), (1,1) and (2,2) is useless, while one having points at (0,0), (2,0), and (0,2) is appropriate. This concept applies as well to systems with more variables. The use of experimental design in multivariate calibration has been discussed in detail elsewhere[16-17].

In measuring the calibration set, randomization is useful in cases where some sort of time dependent phenomena occurs. For modelling the calibration set by regression, models are sought which describe the data well (high r-squared) and have little or no lack-of-fit. Deming and Morgan[15] have described how analysis of variance (ANOVA) may be applied in distinguishing lack-of-fit and purely experimental uncertainty situations. An additional useful check of an ICP-MS calibration model is to determine if the obtained regression coefficients (*i.e.* a and b in Equations 1 and 2) are reasonably consistent with the isotopic ratios.

For analyzing and predicting unknowns, one should be concerned with the possibility that the sample may contain some source of signal not accounted for in the calibration set. Ideally, this issue is resolved at Step 1. PLSR addresses this question on a sample-by-sample basis; PCR and MLR are blind to this possibility. Usual quality control measures such as verification of the calibration using independently obtained standards, verifying continuing calibration, and the analysis of replicates, spiked samples, and reference materials are applicable when using multivariate as well as univariate calibration algorithms.

3 EXPERIMENTAL

Deionized, distilled water was used as the solvent for all solutions. Trace-metal grade nitric and hydrofluoric acids, 70 and 48 wt %, respectively, were used as received from commercial sources. Reagent grade potassium hydroxide was used without any further purification. Stock solutions of analytes and interferants were used as received from Spex Industries (Edison, NJ, USA). The following reference materials were obtained from the US National Institute of Standards and Technology (formerly National Bureau of Standards): SRM 688 (Basalt Rock), SRM 1646 (Estuarine Sediment), SRM 69b (Arkansas Bauxite) and 120b (Florida Phosphate Rock). Reference Soil SO-2 (B-Horizon, Ferro-Humic Podzol) was obtained from the Canadian Certified Reference Materials Project (Energy, Mines and Resources Canada).

Potassium hydroxide fusions were accomplished by mixing

about 0.25 g of sample with 2.0 ± 0.1 g of KOH in a 10 mL pyrolytic graphite crucible, followed by stepwise heating at 160°, 300°, 400° and 475° C for one hour at each temperature. The melts were then dissolved with 40 mL of deionized water containing 10 mL of 70 weight percent aqueous nitric acid. This step was followed by filtration through a 0.45 micron membrane filter and dilution to 100 mL with deionized water. Complete dissolution was evident in most cases. These dissolved samples were generally diluted 25-fold with water prior to ICP-MS analysis. Reagent blanks were prepared at a minimum of three per batch.

Nitric-hydrofluoric acid microwave assisted digestions were performed using a Model MDS-205 sample preparation system (CEM Corporation, Matthews, NC, USA). A 1 g portion of sample, 10 mL of 70 % HNO_3, and 10 mL of 48 % HF were mixed in a PTFE pressure-relief vessel and heated for 50 minutes at 360 watts (based on a 12 vessel batch). Venting did not take place under these conditions. A small amount of a white precipitate (presumably silica) was evident upon filtration through a 0.45 micron membrane filter.

Table 1 Operating parameters for ICP-MS

Forward RF power	1250 W
Reflected power	5 W
Coolant Ar flow	15.0 $L.min^{-1}$
Auxiliary Ar flow	1.4 $L.min^{-1}$
Nebulizer Ar flow	1.0-1.2 $L.min^{-1}$
Sample uptake rate	1.0-2.0 $mL.min^{-1}$
Sampling depth	25 mm
Mass spectrometer pressure	1.5-5.0 x 10^{-5} torr
Interface pressure	1.0-1.5 torr
Electron multiplier voltage	3930 V
Deflector voltage	4030 V
B lens setting	30
E_1 lens setting	42
P lens setting	17
S_2 lens setting	20
Resolution	low (1.0-1.1 amu at 10% height)
Measurement mode	multichannel
Measurement time	5-10 seconds
Dwell time	0.050 seconds
Cycle time	0.100 seconds
Measurements/peak	1
Repeats/integration	1-3
TiO^+/Ti^+	0.005-0.010

The filtered digests were diluted to 100 mL with deionized water, and were further diluted 20-fold with water prior to ICP-MS analysis. Reagent blanks were prepared at a minimum of three per batch.

A Sciex Elan Model 250 ICP-MS, equipped with mass flow meters and a peristaltic sample delivery pump was used in these studies. A refrigerated circulating bath was used to maintain the spray-chamber temperature at 10° centigrade. Meinhard type TR-C concentric glass nebulizers were used in all studies. Typical operating conditions were as shown in Table 1.

The Ti-Zn system was surveyed using experiments where a HNO_3/HF preparation of NIST 688 was analyzed with sixteen additions using the Plackett-Burman design shown in Table 2. In this experiment, masses 64 and 66 were scanned as described in Table 1. The resulting signals were normalized to that of $^{69}Ga^+$ which was used as an internal standard. This experiment neglected the effects of signals from adjacent masses not resolved by the quadrupole mass analyzer (*i.e.* Cu was not included in the design) and the effect of $^{64}Ni^+$ (0.95% abundance), which had been previously established to be negligible for these samples. The unused variables, error 1 through error 5, constitute degrees of freedom for the experimental design; variance which appears to arise from these pseudo-variables is indicative of specific interactions of variables, or a high level of purely experimental uncertainty of the system. Calibrations were conducted using GSAM and external calibration (EXTCAL), both followed by multiple linear regression treatment of the data. Pertinent information is summarized in Table 3. $^{69}Ga^+$, $^{103}Rh^+$, and $^{89}Y^+$ were used as internal standards for GSAM-I, EXTCAL-I, and EXTCAL-II, respectively. Calibration designs used in these studies are listed in Table 4. The data obtained from the calibration designs shown in Table 4 were analyzed by multiple linear regression using a commercially available statistics software package (Statgraphics, STSC Inc., Rockville, MD, USA). The internal standard-normalized intensities were treated as dependent variables; the concentrations of Ti, Cu and Zn were independent variables. At 63 and 65 amu, Ti and Cu concentrations were considered, while 64 and 66 amu were modelled using Ti and Zn. GSAM involved writing equations in terms of the added concentrations of Ti, Cu, and Zn, while the concentrations of external standards were used in the EXTCAL processes. The GSAM equations were solved using the mean response values of the reagent blanks at each mass, while for EXTCAL, the contribution of reagent blanks were discretely determined and their mean was subtracted from the sample result.

4 RESULTS AND DISCUSSION

The results of the Plackett-Burman screening experiment (Table 2) were investigated using multiple linear regression. In short, a "model" was constructed which attempted to explain the observed variances in the signals at amu 64 and 66 in terms of the concentrations of species added (see Table 2). The statistical significance of each addition was determined by calculating the two-sample t-statistic for its addition,

Table 2 Plackett-Burman survey of signals at 64 and 66 amu.

Variable	Species	Addition 1	Addition 2
1	Cl	0	40
2	Si	0	40
3	error 1	-	-
4	Al	0	40
5	Na	0	40
6	error 2	-	-
7	Mg	0	40
8	P	0	40
9	S	0	40
10	Ca	0	40
11	Ti	0	4
12	Zn	0	0.04
13	error 3	-	-
14	error 4	-	-
15	error 5	-	-

Additions 1 and 2 in $\mu g.mL^{-1}$

Addition	Variable number														
	1	2	3	4	5	6	7	8	9	10	11	12	13	14	15
1	1	1	1	1	1	1	1	1	1	1	1	1	1	1	1
2	1	1	1	1	1	1	1	2	2	2	2	2	2	2	2
3	1	1	1	2	2	2	2	1	1	1	1	2	2	2	2
4	1	1	1	2	2	2	2	2	2	2	2	1	1	1	1
5	1	2	2	1	1	2	2	1	1	2	2	1	1	2	2
6	1	2	2	1	1	2	2	2	2	1	1	2	2	1	1
7	1	2	2	2	2	1	1	1	1	2	2	2	2	1	1
8	1	2	2	2	2	1	1	2	2	1	1	1	1	2	2
9	2	1	2	1	2	1	2	1	2	1	2	1	2	1	2
10	2	1	2	1	2	1	2	2	1	2	1	2	1	2	1
11	2	1	2	2	1	2	1	1	2	1	2	2	1	2	1
12	2	1	2	2	1	2	1	2	1	2	1	1	2	1	2
13	2	2	1	1	2	2	1	1	2	2	1	1	2	2	1
14	2	2	1	1	2	2	1	2	1	1	2	2	1	1	2
15	2	2	1	2	1	1	2	1	2	2	1	2	1	1	2
16	2	2	1	2	1	1	2	2	1	1	2	1	2	2	1

Table 3 Calibration algorithms for the determination of Ti, Cu and Zn in reference materials by ICP-MS with multivariate calibration.

Design	Sample, Preparation	Analytes	Sensors
GSAM-I	NIST 688, KOH	Ti, Cu, Zn	63-66
GSAM-I	SO-2, KOH	Ti, Cu, Zn	63-66
EXTCAL-I	NIST 120b, KOH	Ti, Cu, Zn	63-66
EXTCAL-I	NIST 69b, KOH	Ti, Cu, Zn	63-66
EXTCAL-I	NIST 1646, KOH	Ti, Cu, Zn	63-66
EXTCAL-I	NIST 688, KOH	Ti, Cu, Zn	63-66
EXTCAL-II	NIST 688, HNO_3/HF	Ti, Zn	64,66

which compares the eight samples to which it was added with those for which no addition was made. For NIST 688 at 64 amu, the results of this experiment are shown in Table 5. This table shows the regression coefficients (the effect of a 1 $\mu g.mL^{-1}$ addition of the analyte upon the 64-69 amu ratio), the two-sample t-statistic for the regression coefficient, and the significance level (expressed as the probability that the regression coefficient is different from zero). The additions of Ti and Zn to this sample by themselves accounted for 81.88% of the dataset's variance, while the entire model shown in Table 5 explained 88.03% of the variance. It is evident from examination of Table 5 that Cl and possibly Mg produce some enhancement of the signal at 64 amu, however, their effects are small on a concentration basis (a 1 $\mu g.mL^{-1}$ addition to this preparation represents a 2000 $\mu g.g^{-1}$ "spike" of the original sample). High levels of Al, Na, and S appear to suppress the internal standard-normalized signals. Effects which possibly explain the variance not accounted for by the model are: (a) drift in the TiO^+/Ti^+ ratio, (b) failure of the internal standardization process to compensate for random changes of analyte sensitivity, (c) interactions of the main effects, and (d) purely experimental uncertainty.

The results obtained for the same analysis at amu 66 closely paralleled those of Table 5. Based upon this exercise, the sample-derived signals at masses 64 and 66 were treated as being strictly due to Ti and Zn. Table 6 depicts results for the determination of Ti, Cu and Zn in the reference materials using GSAM and EXTCAL followed by multiple linear regression. In general, the results of Table 6 indicate that all calibration schemes are successful for these determinations. With perhaps the exception of determinations of Ti in NIST 1646, the precision of these results is typical of that seen for determinations by ICP-MS with univariate calibration.

Table 4 Calibration designs used for quantitative analysis

A GSAM-I

Addition	Ti added	Cu added	Zn added
0	0	0	0
1	0	0.01	0.01
2	0	0.02	0.02
3	0	0.05	0.05
4	1	0	0
5	2	0	0
6	5	0	0

B EXTCAL-I

Standard	Ti conc.	Cu conc.	Zn conc.
1	0	0	0
2	0	0.02	0.02
3	2	0	0.02
4	2	0.02	0

C EXTCAL-II

Standard	Ti conc.	Zn conc.
1	0	0
2	0	0.04
3	0	0.08
4	2	0
5	2	0.04
6	2	0.08
7	4	0
8	4	0.04
9	4	0.08

Elemental concentrations in $\mu g.mL^{-1}$

Both the GSAM and external calibration schemes are capable of producing satisfactory analytical results for analytes such as Cu and Zn under circumstances where an interfering polyatomic species is a major contributor to the analytical signal. A previous study[14] found that cadmium could be determined in the presence of MoO^+ using MLR and PCR, under conditions where the latter species were responsible for up to ten times the signal level of Cd. Precisions and biases for the Cd results were less than 10 relative percent.

Table 5 Effects of the addition of possible interferants upon NIST 688 signals at 64 amu

Species	Coefficient	t-value	Significance level
Ti	2.6E-3	16.09	0.9999+
Zn	0.147	8.05	0.9999+
Al	-5.1E-5	-3.39	0.9986
Cl	4.9E-5	3.13	0.9970
S	-4.6E-5	-2.69	0.9900
Na	-4.5E-5	-2.45	0.9817
Mg	4.5E-5	2.14	0.9620
Ca	-1.9E-5	-1.25	0.7924
P	6.8E-6	0.44	0.3379
Si	3.3E-8	0.002	0.0017

Some post-digestion spike recovery experiments were also performed upon diluted aliquots of a KOH-fused NIST 1646 sample. These experiments had the dual purpose of demonstrating that the calibration system could properly measure the analyte when spiked, and that it would not falsely measure an analyte when it was not spiked. These results are shown in Table 7. The spike levels added reflect those calculated in terms of the original sample. Clearly, the measurement system is robust with respect to distinguishing the source of signal. These post-digestion spike recovery results indicate that the GSAM and EXTCAL algorithms should be essentially equivalent; this is, in fact, evident from the sample results of Table 6.

5 CONCLUSIONS

It has been shown that multiple linear regression is a useful calibration method for the ICP-MS determination of analytes in the presence of interferences. MLR can either be implemented in a standard addition (GSAM) or external calibration format. A variety of calibration designs can be used to set up calibration sets which adequately describe the effects of signal sources at the sensors of interest. MLR as well as other multi-variate calibration approaches not considered herein appear to hold promise for the determination of analytes in real samples. This work has made no attempt to correct for the effects of possible drift in the MO^+/M^+ ratio; such drift has been observed by others[24] but did not cause any large problems herein.

Table 6 Results for the determination of Ti, Cu, and Zn in reference materials

Material	Method	Certified values		Number of replicates	Results
NIST 688	KOH	Ti	7010	3	6900 ± 500
	GSAM-I	Cu	(96)	3	92 ± 4
		Zn	(58)	3	71 ± 3
SO-2	KOH	Ti	8600	1	8930
	GSAM-I	Cu	7	1	(<10)
		Zn	124	1	125
NIST 120b	KOH	Ti	900	2	1330 ± 40
	EXTCAL-I	Cu	NI	2	18 ± 3
		Zn	NI	2	124 ± 4
NIST 69b	KOH	Ti	11400	2	11600 ± 100
	EXTCAL-I	Cu	NI	2	11 ± 3
		Zn	28	2	25 ± 6
NIST 688	KOH	Ti	7010	2	6300 ± 300
	EXTCAL-I	Cu	(96)	2	87 ± 7
		Zn	(58)	2	70 ± 10
NIST 1646	KOH	Ti	5100	2	5000 ± 1000
	EXTCAL-I	Cu	18	2	23 ± 3
		Zn	138	2	126 ± 6
NIST 688	HNO_3/HF	Ti	7010	5	8000 ± 300
	EXTCAL-II	Zn	(58)	5	67 ± 1

Explanatory notes: All concentrations are in $\mu g.g^{-1}$. All replicates are complete preparations and measurements. Uncertainties refer to one standard deviation for 3 or 5 measurements and to the range for two measurements. The Cu detection limit was estimated from the precision information of the other samples. The Cu and Zn values in parentheses are not certified, and are quoted by NIST for "information purposes". "NI" indicates that no information is provided by NIST about this element's content.

Plackett-Burman screening experiments are of use for rapidly evaluating the potential sources of signal at a given sensor. These experiments provide perhaps the most time-effective way for the analyst to investigate any physical system by simultaneous manipulation of several variables. Plackett-Burman experiments can be used to determine if multivariate calibration of any sort (or some other interference management mechanism) is required, and to assist in construction of suitable multivariate calibration systems.

Table 7 Spike recovery results for post-digestion spikes of NIST 1646 using the EXTCAL-I algorithm

Native concentrations found: Ti 5000 ± 1000; Cu 23 ± 3; Zn 126 ± 6

Experiment	Spike levels		Spiked sample results
1	Ti	9900	14500
	Cu	0	23
	Zn	0	124
2	Ti	0	5100
	Cu	198	223
	Zn	0	126
3	Ti	0	4900
	Cu	0	23
	Zn	396	540
4	Ti	19800	25500
	Cu	0	20
	Zn	792	934

Concentrations in $\mu g.mL^{-1}$

A wide variety of isobaric polyatomic interference problems exist in ICP-MS; many of them are potentially manageable using multivariate calibration. Compromises in sample preparation (*i.e.* avoiding HCl), instrument conditions, or preliminary chemical separations can potentially be avoided with this methodology. One must, however, contend with larger numbers of standards and a more complicated calibration model. These difficulties may be minimized through automated operation and a proper data handling system; both are well within the realm of existing analytical technology. Application of multivariate calibration to other interference problems in ICP-MS continues in our laboratories.

ACKNOWLEDGEMENTS

This work was supported by the U.S. Environmental Protection Agency. Mention of specific products and vendors is for information purposes only and does not constitute official endorsement.

REFERENCES

1. A.R. Date and A.L. Gray, eds. 'Applications of Inductively Coupled Plasma Mass Spectrometry', Blackie, London, 1989.
2. R.S. Houk, H.J. Svec and V.A. Fassel, *Appl. Spec.*, 1981, **35**, 380.
3. D.J. Douglas, E.S.K. Quan and R.G. Smith, *Spectrochim. Acta*, 1983, **38B**, 39.
4. A.L. Gray, *J. Anal. At. Spectrom.*, 1986, **1**, 247.
5. G. Horlick, S.H. Tan, M.A. Vaughn and C.A. Rose, *Spectrochim. Acta*, 1985, **40B**, 1555.
6. R.C. Hutton and A.N. Eaton, *J. Anal. At. Spectrom.*, 1987, **2**, 595.
7. D. Beauchemin, Abstracts of the 3rd Surrey Conference on Plasma Source Mass Spectrometry, July 1989.
8. L.M. Faires, Abstracts of the 3rd Surrey Conference on Plasma Source Mass Spectrometry, July 1989.
9. F.E. Lichte, A.L. Meier and J.G. Crock, *Anal. Chem.*, 1987, **59**, 1150.
10. B.E.H. Saxberg and B.R. Kowalski, *Anal. Chem.*, 1979, **51**, 1031.
11. R.W. Gerlach and B.R. Kowalski, *Anal. Chim. Acta*, 1982, **134**, 119.
12. P.J. Gemperline, S.E. Boyette and K. Tyndall, *Appl. Spec.*, 1987, **41**, 454.
13. K.R. Beebe and B.R. Kowalski, *Anal. Chem.*, 1987, **59**, 1007A.
14. M.E. Ketterer, J.J. Reschl and M.J. Peters, *Anal. Chem.*, 1989, **61**, 2031.
15. S.N. Deming and S.L. Morgan, 'Experimental Design: A Chemometric Approach', Elsevier, Amsterdam, 1987.
16. H. Mark, *Anal. Chem.*, 1986, **58**, 2814.
17. F. Cahn and S. Compton, *Appl. Spec.*, 1988, **42**, 856.

EVALUATION OF ICP-MS FOR THE DETERMINATION OF TRACE AND ULTRA-TRACE ELEMENTS IN HUMAN SERUM AFTER SIMPLE DILUTION.

H. Vanhoe, C. Vandecasteele[1], J. Versieck[2] and R. Dams

Laboratory of Analytical Chemistry, Rijksuniversiteit Gent, Proeftuinstraat 86, B-9000 Gent, Belgium

[1]Research Director of the National Fund for Scientific Research Belgium

[2]Department of Internal Medicine, Division of Gastroenterology, University Hospital, De Pintelaan 185, B-9000 Gent, Belgium

1 INTRODUCTION

In spite of the rapid increase in interest in inductively coupled plasma mass spectrometry (ICP-MS), there are still only a few papers devoted to the analysis of biological materials. These can be divided into two groups, namely those concerned with elemental analysis and those concerned with isotopic analysis. A number of trace elements have been determined in several biological reference materials: National Institute of Standards and Technology (NIST) SRM 1566 Oyster Tissue[1] (V, Mo, Cd, Pb); SRM 1567 WheatFlour[2,3] (Al, Cr, Ni, Zn, Rb, Mo, Cd, Sn, Ce, Pb); SRM 1568 Rice Flour[2] (Co, Ni, As, Mo, Cd, Ce, Pb); SRM 1570 Spinach[3] (Al, Cr, Ni, Zn, Rb, Mo, Cd, Pb); SRM 1572 Citrus Leaves[2] (Ni, Cd, Sn, La, Ce, Tl); SRM 1577a Bovine Liver[1,4,5,6] (Na, Mg, P, S, K, Ca, V, Mn, Fe, Co, Cu, Zn, Se, Rb, Sr, Mo, Cd, Pb); National Research Council of Canada (NRCC) TORT-1 Lobster Hepatopancreas[7] (V, Cr, Mn, Fe, Ni, Co, Cu, Zn, As, Se, Sr, Mo, Cd, Sn, Hg, Pb); DORM-1 Dogfish Muscle Tissue[8] and DOLT-1 Dogfish Liver Tissue[8] (Na, Mg, Cr, Fe, Mn, Co, Ni, Cu, Zn, As, Cd, Hg, Pb); NOAAK Cod Liver Tissue[9] and NOAAL Shellfish Tissue[9] (Cr, Fe, Mn, Ni, Cu, Zn, As, Ag, Cd, Sn, Hg, Pb). The isotope ratios of Li[10], Mg[11], Fe[12,13], Cu[12], Zn[12,14], Se[6], Br[15], and Pb[16,17] have been investigated.

The purpose of this paper is to investigate the suitability of ICP-MS for the determination of trace elements in one of the most important biological materials from a clinical point of view, namely human serum. A part of this work was described earlier in more detail by Vanhoe *et al.*[18] and by Vandecasteele *et al.*[19]. It is part of an ongoing project, results of which will be reported in due course. Several attempts to analyse serum have been described in the literature. Some of these dealt with isotope ratio determinations of

Fe[13], Zn[14], Se[6], Br[15]. Lyon *et al.*[20] performed a chromatographic separation before measuring selenium, and also measured several elements (Al, V, Cr, Mn, Fe, Ni, Cu, Zn, Se, Mo, and Ba) in a plasma protein solution without prior separation[5]. However, the concentration levels for various elements were much higher than those in real serum samples. Our aim was to determine trace elements in uncontaminated serum samples. Therefore, a "second-generation biological reference material (human serum)" prepared by Versieck *et al.*[21], was used. This material was prepared under rigorously controlled conditions to avoid contamination. The trace-element concentrations in the serum approximate those in normal human serum. Our results were based on the analysis of liquid serum. The reference material is, however, available to the scientific community in freeze-dried form. Experiments showed that reconstitution of the freeze-dried human serum seemed not to be a problem, and comparative studies are underway. In addition, it was considered of prime importance to limit sample pretreatment as much as possible. A five-fold dilution of the serum samples with 0.14 M nitric acid was therefore the only sample pretreatment. This dilution is necessary to avoid rapid blockage of the central tube of the plasma torch and to reduce the extent of signal suppression due to easily ionised elements, *e.g.* sodium, present in the sample. Vandecasteele *et al.*[22] showed that using indium as an internal standard corrects for this matrix effect.

2 EXPERIMENTAL

Instrumentation

A commercially available ICP-MS instrument, the VG PlasmaQuad (VG Elemental, Winsford, U.K.), was used in its standard configuration with a Meinhard nebuliser and a double pass, water-cooled spray chamber made of borosilicate glass. Details of the operating conditions are given in Table 1.

Reagents and standards

High purity water used in this work, was obtained with a Millipore Milli-Q water purification system (18 MΩ.cm^{-1} specific resistivity). Concentrated nitric acid (65%) was prepared by sub-boiling distillation in a quartz still using analytical-reagent grade nitric acid (Pleuger) as feedstock. This ultrapure acid was used throughout this work.

For the quantitative analysis various multi-element standard solutions (100 μg.L^{-1} of the analyte elements except for Br where 1 mg.L^{-1}) were used. Prior to each analysis working standards were prepared from multi-element stock solutions (10 mg.L^{-1}) by successive

Table 1 ICP-MS instrument operating conditions

Instrument	VG Elemental PlasmaQuad
forward power	1.35 kW
reflected power	<10 W
plasma gas	13 $L.min^{-1}$
auxiliary gas	0.8 $L.min^{-1}$
nebuliser gas	0.73 $L.min^{-1}$
peristaltic pump	Gilson Minipuls 2, pumped at 0.9 $mL.min^{-1}$
nebuliser	Meinhard concentric glass nebuliser (type: Tr-30-A3)
spray chamber	Scott double pass, water cooled (10°C)
Ion sampling	
sampling cone	nickel; 1.0mm orifice
skimmer cone	nickel; 0.75mm orifice
sampling depth	10mm from load coil
Vacuum	
expansion	2.3 mbar
intermediate	$1.0x10^{-4}$ mbar
analyser	$4.0x10^{-6}$ mbar

dilutions with 0.14 M nitric acid. Nine stock solutions were prepared: Li, Na, K, Rb, and Cs; Be, Mg, Ca, Sr, Ba, La, and Y; B, Al, Ga, Cd, Tl, and Pb; Ge, As, Se, Sn, Sb, and Te; V, Cr, Mn, Fe, Co, Ni, Cu, and Zn; Zr, Nb, Mo, Hf, Ta, and W; Br; Hg. They all contained 0.70 M nitric acid, except for the solution containing Mo which also contained 0.58 M hydrofluoric acid (reagent grade, UCB) and for the solution of Hg which also contained 0.025% potassium dichromate. All these stock solutions were prepared from commercially available trace-element solutions (1 $g.L^{-1}$), except for the Hg solution which was prepared in our laboratory[23], and were stored in polyethylene flasks. An indium solution (as internal standard) was added to the final multi-element standard solutions (100 $\mu g.L^{-1}$, for Br 1 $mg.L^{-1}$) to obtain an indium concentration of 100 $\mu g.L^{-1}$.

Sample preparation

Liquid serum, obtained by defrosting serum that had been stored in a polyethylene container in a deep freezer, was diluted five-fold with 0.14 M nitric acid. To 5 mL of serum, 2.5 mL of a 1 $mg.L^{-1}$ indium solution was added and the volume was adjusted with 0.14 M nitric acid to obtain 25 mL of the serum solution with an indium concentration of 100 $\mu g.L^{-1}$. Precleaned polyethylene pipettes and volumetric flasks were used to prepare the diluted serum samples and all manipulations were carried out on a clean-bench to avoid contamination.

Blank solution

In order to correct for interferences due to the formation of polyatomic ions, a simulated blank solution was prepared for the determination of iron, cobalt, copper, and zinc in human serum. This solution contained the following components: 0.14 M nitric acid, internal standard (100 $\mu g.L^{-1}$ of indium), sodium chloride (reagent grade, UCB), sodium nitrate (reagent grade, Carlo Erba), cysteine (cysteine chloride monohydrate, reagent grade, Merck), and calcium nitrate (calcium nitrate tetrahydrate, reagent grade, Merck). The latter four components yield the same concentration of sodium, chlorine, sulphur, and calcium as in human serum. The corresponding concentrations are given in Table 2. 0.14 M nitric acid with 100 $\mu g.L^{-1}$ of indium as internal standard was used as blank for the determination of all the other elements and for the multi-element standard solutions.

Table 2 Composition of the simulated blank solution for the determination of Fe, Co, Cu, and Zn in a five-fold diluted serum solution

1.2 $g.L^{-1}$ NaCl	(730 $mg.L^{-1}$ Cl & 470 $mg.L^{-1}$ Na)
0.65 $g.L^{-1}$ $NaNO_3$	(180 $mg.L^{-1}$ Na)
1.3 $g.L^{-1}$ cysteine.HCl.H_2O	(240 $mg.L^{-1}$ S)
0.11 $g.L^{-1}$ $Ca(NO_3)_2.4H_2O$	(19 $mg.L^{-1}$ Ca)
100 $\mu g.L^{-1}$ In	(internal standard)
0.14 M nitric acid	(solvent)

Analytical procedure

Instrument operating conditions Each time an analysis was carried out, the ion lens voltages of the instrument and the plasma operating conditions were re-adjusted resulting in a compromise between the sensitivity and the oxide levels. Additionally, before each analysis sequence a mass calibration was performed with a multi-element standard solution (100 $\mu g.L^{-1}$ of Be, Al, Co, Rb, In, La, and U) to obtain a perfect channel to mass correlation.

Measurements The scanning mode for data acquisition was used for the quantitative analyses. The mass range selected depended on the elements considered. At least 20 channels per mass were used and a measuring time of one minute was usually sufficient.

In order to reduce memory effects for some elements[18] the following analysis sequence was applied : first a blank for the serum samples was measured, then two diluted serum solutions, then a blank solution for the standards, and only at the end of the sequence multi-element standard solutions were measured. Three replicate measurements were made on each solution and this sequence was repeated a second time the same day. A few days later, the same analysis was carried out on two freshly diluted serum samples. After measurement of each solution, the sample introduction system was rinsed for at least three minutes with 0.14 M nitric acid. As an additional precaution the skimmer, the sampling cone, the plasma torch, and the spray chamber were cleaned thoroughly each time before an analysis sequence was carried out.

Calculations For each solution (blank, sample, standard), the signal peak areas integrated over 0.8 amu around the peak maximum of the analyte elements were divided (normalised) by the signal of the internal standard (^{115}In). The mean and standard deviation of the three resulting normalised signals of each solution were calculated. For each element the average normalised signal of the blank was subtracted from that of the serum solutions. External calibration, using multi-element standards, was employed to calculate the corresponding concentrations.

3 RESULTS AND DISCUSSION

Determination of the first row transition metals and of aluminium, arsenic, and selenium

For the determination of elements with a mass below 80, potential interferences from polyatomic ions must be considered[24]. When biological materials are analysed, such interferences originate mainly from argon, nitrogen, and/or oxygen in combination with carbon, sodium, sulphur, chlorine, and calcium, which are present respectively at concentrations of about 35000, 3251 (3130 - 3370), 1197 (1120 - 1270), 3655 (2940 - 4120), and 97 $mg.L^{-1}$ (92 - 109 $mg.L^{-1}$) in human serum[25]. The values in parentheses give the range due to the natural variation of the element considered.

Without special treatment of the serum samples, in order to remove or reduce the sources of polyatomic interferences, and using conventional pneumatic nebulisation for sample introduction, the determination of Al, Cr, Mn, As, and Se is excluded, since the apparent concentration exceeds, even for the most suitable isotopes, by far the real concentration. Applying a suitable correction for the polyatomic interferences will prohibitively enlarge the statistical error on the final result. It is also not possible to determine without any special sample pretreatment titanium, vanadium and nickel, three other transition metals for the same reason.

Table 3 Apparent concentration (μg.L^{-1}) of Fe, Co, Cu, and Zn in a simulated blank solution containing the same concentration of Na, Cl, S, and Ca as a five-fold diluted serum solution

Nuclide (%)	Polyatomic interferences	Simulated blank	Conc. in human serum**
^{57}Fe (2.19)	^{40}Ar^{16}OH,^{40}Ca^{16}OH	330 ± 27*	2350 ± 140
^{59}Co (100)	^{36}Ar^{23}Na, ^{43}Ca^{16}O, ^{40}Ar^{18}OH	0.08 ± 0.02	0.33 ± 0.05
^{63}Cu (69.1)	^{40}Ar^{23}Na	420 ± 50	1009 ± 36
^{65}Cu (30.9)	^{33}S^{16}O^{16}O,^{32}S^{16}O^{17}O	8.2 ± 1.0	
^{64}Zn (48.9)	^{32}S^{16}O^{16}O	62 ± 6	873 ± 1

* 95% confidence limits.
** reference 21, this concentration (μg.L^{-1}) must be divided by 5 for comparison with the values in column 3.

Table 4 Comparison of the results (μg.L^{-1}) for the second-generation biological reference material (human serum)

Element	Isotope	ICP-MS	Certified value
iron	^{57}Fe	2210 ± 350*	2350 ± 140
cobalt	^{59}Co	0.24 ± 0.05	0.33 ± 0.05
copper	^{65}Cu	1047 ± 52	973 ± 36
zinc	^{66}Zn	939 ± 21	873 ± 18
bromine	^{79}Br	4280 ± 160	4440 ± 350
rubidium	^{85}Rb	159.8 ± 1.0	168 ± 30
molybdenum	^{98}Mo	0.67 ± 0.08	0.68 ± 0.07
cadmium	^{111}Cd	< 0.7	0.155 - 0.227
cesium	^{133}Cs	0.92 ± 0.05	0.91 ± 0.21
mercury	^{202}Hg	< 1.7	0.56 - 0.64**

* ± values for ICP-MS are 95% confidence limits.
** reference 29.

For the other first row transition metals (Fe, Co, Cu, and Zn) accurate results were obtained by using a simulated blank solution in order to correct for polyatomic interferences. This solution contains the same concentration of sodium, sulphur, chlorine, and calcium as the five-fold diluted serum samples as described in the experimental section and in more detail in reference 18.

Table 3 gives the contribution of the polyatomic interferences to the total signal in human serum for some selected isotopes of Fe, Co, Cu, and Zn, and Table 4 shows the results obtained by ICP-MS for the reference serum.

Iron Determinations were made at mass 57 (^{57}Fe, 2.19% abundant) to avoid $^{40}Ar^{16}O^+$, $^{40}Ca^{16}O^+$, $^{40}Ar^{14}N^+$, $^{40}Ca^{14}N^+$, and $^{37}Cl^{16}OH^+$ at mass 56 (^{56}Fe, 91.7%) and mass 54 (^{54}Fe, 5.8%). The low precision of the iron results is due to the large interference correction required (Table 3). This correction is necessary as ^{57}Fe is interfered by $^{40}Ar^{16}OH^+$ and $^{40}Ca^{16}OH^+$.

Cobalt Cobalt is mono-isotopic (^{59}Co, 100%). The results obtained by ICP-MS are somewhat imprecise because of the low concentration of cobalt in human serum ($< 0.4\ \mu g.L^{-1}$) (Table 3). In addition ^{59}Co is interfered by $^{36}Ar^{23}Na^+$ and $^{43}Ca^{16}O^+$ and therefore a large interference correction is required.

Copper The ICP-MS results were obtained at mass 65 rather than at mass 63 where the species $^{40}Ar^{23}Na^+$ causes a problem, as shown in Table 3. The polyatomic ions interfering at mass 65 ($^{33}S^{16}O^{16}O^+$, $^{32}S^{16}O^{17}O^+$) do not constitute a major problem. The precision of the copper results obtained by ICP-MS is comparable to the uncertainty on the certified value.

Zinc All of the zinc isotopes (^{64}Zn, 48.9%; ^{66}Zn, 27.8%; ^{67}Zn, 4.1%; ^{68}Zn, 18.6%; ^{70}Zn, 0.62%) are interfered to some extent by polyatomic ions containing oxygen and sulphur. Determinations were made at mass 66 as the more abundant ^{64}Zn exhibits a larger interference (Table 3). The precision of the zinc results is comparable to the uncertainty on the certified value.

In general, the results for iron, cobalt, copper, and zinc agree well within the experimental uncertainty of the certified values.

Determination of Br, Rb, Mo, Cd, Cs, and Hg

For elements with a mass above mass 80, polyatomic interferences are much less frequent than at lower masses. Table 4 gives the results for Br, Rb, Mo, Cd, Cs, and Hg obtained thus far by ICP-MS for the reference serum.

Bromine Bromine has two isotopes : ^{79}Br (50.7%) and ^{81}Br

(49.3% abundant). No polyatomic interferences could be observed for the two isotopes, so both are useful for the determinations. Nevertheless, bromine originally presented some problems. Both isotopes are situated in the mass spectrum adjacent to the more intense $^{40}Ar^{40}Ar^{+}$ peak as shown in Figure 1. This figure shows that for the resolution normally used the peaks are not fully separated. The width of the peak at 5% of the peak maximum is about 0.95 amu.

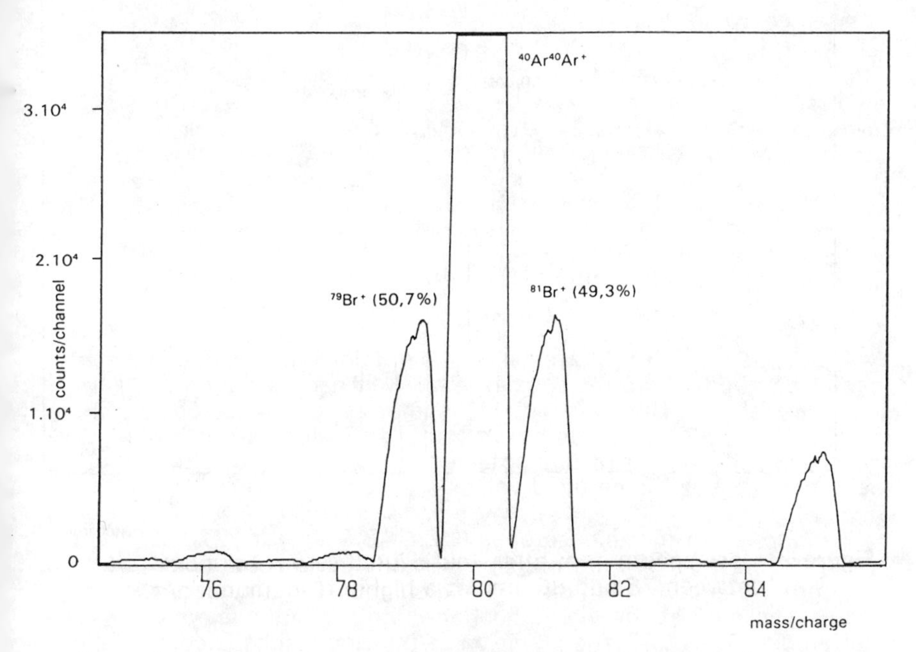

Figure 1 Mass spectrum of a ten-fold diluted serum solution between 75 and 86 amu at normal resolution

The result is that the setting of the peak integration range and mass calibration become critical. When, however, the resolution is improved somewhat (to 0.61 amu), which is quite an easy adjustment to make with a quadrupole mass spectrometer the peaks are clearly separated as shown in Figure 2. Determinations were made at masses 79 and 81. The results were in excellent agreement with each other and with the certified value.

Rubidium Determinations were made at 85 amu (^{85}Rb, 72.2%)

because of an isobaric interference on ^{87}Rb (27.8%) by ^{87}Sr (7.0%). The precision of the rubidium results is significantly smaller than the uncertainty on the certified value.

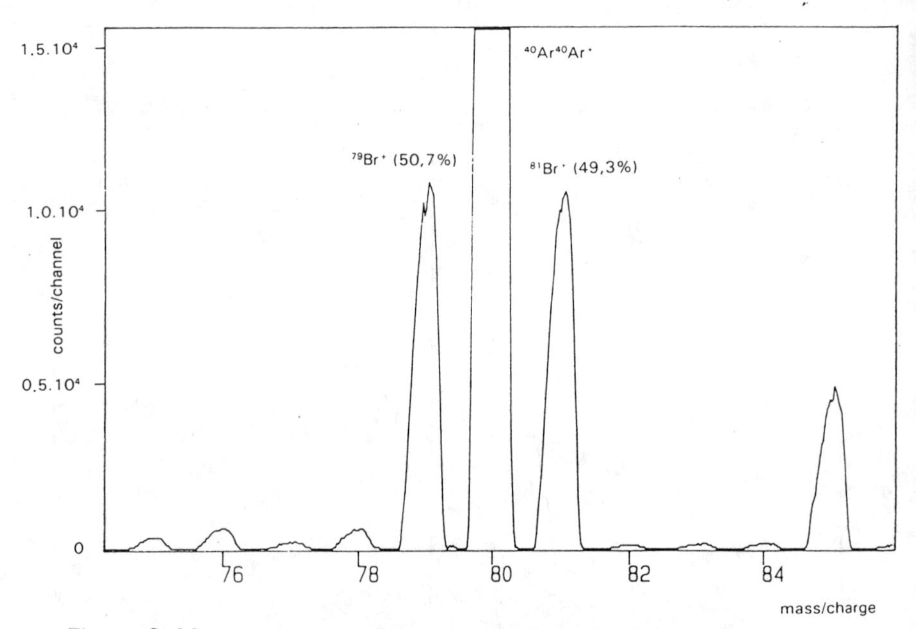

Figure 2 Mass spectrum of a ten-fold diluted serum solution between 75 and 86 amu at a higher resolution

Molybdenum Molybdenum has seven isotopes : ^{92}Mo (14.8%), ^{94}Mo (9.1%), ^{95}Mo (15.9%), ^{96}Mo (16.7%), ^{97}Mo (9.5%), ^{98}Mo (24.4%), and ^{100}Mo (9.6%). Some of these are interfered by polyatomic ions containing bromine and oxygen. Only $^{79}Br^{16}O^+$, interfering at mass 95 and $^{81}Br^{16}O^+$, interfering at mass 97, are important as polyatomic interferences for the considered concentration of bromine[18] (*ca.* 4.4 $mg.L^{-1}$ for the reference serum)[21]. Therefore, determinations were made at mass 98. The precision obtained for molybdenum by ICP-MS is comparable to the uncertainty on the certified value.

Cadmium Cadmium has eight isotopes : ^{106}Cd (1.2%), ^{108}Cd (0.9%), ^{110}Cd (12.4%), ^{111}Cd (12.8%), ^{112}Cd (24.0%), ^{113}Cd

(12.3%), ^{114}Cd (28.8%), and ^{116}Cd (7.6%). There are isobaric interferences on ^{112}Cd, ^{113}Cd, ^{114}Cd, and ^{116}Cd respectively by ^{112}Sn (1.0%), ^{113}In (4.3%, internal standard), ^{114}Sn (0.65%), and ^{116}Sn (14.4%). Determinations were therefore made at masses 110 and 111. Thus far, by using ^{111}Cd and scanning mode for data acquisition, only a detection limit of 0.7 $\mu g.L^{-1}$ could be obtained. The cadmium concentration in the reference serum[21] is *ca.* 0.18 $\mu g.L^{-1}$. Use of the peak jumping mode would probably improve the detection limit for cadmium[26].

Cesium Cesium is mono-isotopic (^{133}Cs, 100%). The precision of the cesium results obtained by ICP-MS is significantly lower than the uncertainty on the certified value.

Mercury Determinations were made at mass 202 (^{202}Hg, 29.7% - the most abundant isotope of mercury). Thus far, only a detection limit of 1.7 $\mu g.L^{-1}$ could be obtained for mercury because of the high blank level which corresponds to *ca.* 0.48 $\mu g.L^{-1}$ of mercury.

In general, the results for Br, Rb, Mo, and Cs obtained by ICP-MS are in good agreement within the experimental uncertainty of the certified values.

Determination of Li and Sr

In addition to evaluating the proposed method for the determination of trace and ultra-trace elements which were already certified in the reference serum, a study was started to determine other ultra-trace elements. Only the results for lithium and strontium will be discussed in some detail. For the other elements, only the possibilty will be mentioned.

Lithium Lithium has two isotopes : ^{6}Li (7.5%) and ^{7}Li (92.5%). Determinations were made at mass 7 as there are no polyatomic interferences reported in the literature[10]. At first instance, the peak jumping mode (^{7}Li, ^{115}In (internal standard) in combination with other nuclides) for data acquisition was used. The results are given in Table 5 which compares these with a value obtained by atomic absorption spectrometry (AAS)[28]. The agreement is not total, but fair in view of the low concentration. Further work is now underway using the scanning mode and ^{9}Be as an internal standard.

Strontium Strontium has four isotopes : ^{84}Sr (0.56%), ^{86}Sr (9.9%), ^{87}Sr (7.0%), and ^{88}Sr (82.6%). Some of these have isobaric interferences: ^{84}Sr, ^{86}Sr, and ^{87}Sr by ^{84}Kr (57.0%), ^{86}Kr (17.3%), and ^{87}Rb (27.8%), respectively. A correction is possible for the first two interferences by using a blank solution, whereas for the third one correction will give less precise results by using the ^{85}Rb (72.2%) nuclide. Determinations were therefore made at mass 88, although using ^{86}Sr did not yield a significant difference[19]. Table 5 gives the results obtained by ICP-MS and compares these with a value obtained by instrumental neutron activation analysis (INAA)[19]. The agreement is good. Results could also be obtained thus far and without special pretreatment of the sample for the elements B, Sn, I, Ba, Pb, and Bi, but these were not yet quantitative. Further work is underway in order to lower the blank levels for B, Sn, and Pb, and to check the accuracy of the method for I, Ba, and Bi.

Table 5 Results for Li and Sr ($\mu g.L^{-1}$) in the second-generation biological reference material (human serum)

Method	Element	
	Li	Sr
ICP-MS*	1.40 ± 0.05**	25.5 ± 0.6**
AAS[28]	1.75 ± 0.05	ND
INAA[19]	ND	22.2 ± 2.1

* this work. ** standard deviation. ND = not determined

4 CONCLUSIONS

It is clear that ICP-MS is a particularly useful analytical technique for the determination of trace and ultra-trace elements in human serum. A great number of elements may accurately and very rapidly be determined: Li, Fe, Co, Cu, Zn, Br, Rb, Sr, Mo, and Cs. For the elements Cd and Hg, thus far, only a detection limit could be obtained.

For the elements Al, Ti, V, Cr, Mn, Ni, As, and Se a simple dilution of the serum samples is not sufficient because of the polyatomic interferences. We are convinced that using an electrothermal volatilisation as a sample introduction system[27], instead of pneumatic nebulisation, presents some advantages for several elements. The use of a drying step before the analysis stage will in fact reduce the population of polyatomic ions produced from the water loading of the plasma (ArO) or from major matrix constituents (*e.g.* CaO, SO, SO_2,). The use of an ashing step, combined with the application of a matrix modifier, will allow elimination of other polyatomic interferences such as those originating from NaCl. The result is that the intensity of ClO, ArCl, NaCl, *etc.* can be significantly reduced.

ACKNOWLEDGMENT

Grateful acknowledgment is made to L. Vanballenberghe and A. Wittoek for their contribution in the preparation of the serum samples and to the I.I.K.W. and N.F.W.O. for financial support. The PlasmaQuad was acquired by a grant of the "Fund for Medical Scientific Research (FGWO)".

REFERENCES

1. S.L. Munro, Les Ebdon, and D.J. McWeeny, *J. Anal. At. Spectrom.*, 1986, **1**, 211.
2. A.W. Boorn, J.E. Fulford, and W. Wegscheider, *Mikrochim. Acta*, 1985, **2**, 171.
3. R.D. Satzger, *Anal. Chem.*, 1988, **60**, 2500.
4. J.E. Fulford and B.C. Gale, *Anal. Proc.*, 1987, **24**, 10.
5. T.D.B. Lyon, G.S. Fell, R.C. Hutton, and A.N. Eaton, *J. Anal. At. Spectrom.*, 1988, **3**, 265.
6. B.T.G. Ting, C.S. Mooers, and M. Janghorbani, *Analyst*, 1989, **114**, 667.
7. P.S. Ridout, H.R. Jones, and J.G. Williams, *Analyst*, 1988, **113**, 1383.
8. D. Beauchemin, J.W. McLaren, S.N. Willie, and S.S. Berman, *Anal. Chem.*, 1988, **60**, 687.
9. D. Beauchemin, J.W. McLaren, and S.S. Berman, *J. Anal. At. Spectrom.*, 1988, **3**, 775.
10. X.F. Sun, B.T.G. Ting, S.H. Zeisel, and M. Janghorbani, *Analyst*, 1987, **112**, 1223.
11. S. Schuette, D. Vereault, B.T.G. Ting, and M. Janghorbani, *Analyst*, 1988, **113**, 1837.
12. B.T.G. Ting and M. Janghorbani, *Spectrochim. Acta*, Part B, 1987, **42**, 21.
13. P.G. Whittaker, T. Lind, J.G. Williams, and A.L. Gray, *Analyst*, 1989, **114**, 675.
14. R.E. Serfass, J.J. Thompson, R.S. Houk, *Anal. Chim. Acta*, 1986, **188**, 73.
15. M. Janghorbani, T.A. Davis, and B.T.G. Ting, *Analyst*, 1988, **113**, 405.
16. H.T. Delves and M.J. Campbell, *J. Anal. At. Spectrom.*, 1988, **3**, 343.
17. M.J. Campbell, and H.T. Delves, *J. Anal. At. Spectrom.*, 1989, **4**, 235.
18. H. Vanhoe, C. Vandecasteele, J. Versieck, and R. Dams, *Anal. Chem.*, 1989, **61**, 1851.
19. C. Vandecasteele, H. Vanhoe, L. Vanballenberghe, A. Wittoek, J. Versieck, and R. Dams, *Talanta*, submitted for publication.
20. T.D.B. Lyon, G.S. Fell, R.C. Hutton, and A.N. Eaton, *J. Anal. At. Spectrom.*, 1988, **3**, 601.
21. J. Versieck, L. Vanballenberghe, A. De Kesel, J. Hoste, B.

Wallays, J. Vandenhaute, N. Baeck, and F.W.Jr Sunderman, *Anal. Chim. Acta*, 1988, **204**, 63.

22. C. Vandecasteele, M. Nagels, H. Vanhoe, and R. Dams, *Anal. Chim. Acta*, 1988, **211**, 91.
23. E. Temmerman, Ph.D.Thesis, Ghent University, Belgium, 1988.
24. S.H. Tan, and G. Horlick, *Appl. Spectrosc.*, 1986, **40**, 445.
25. G.V. Iyengar, W.E. Kollmer, and H.J.M. Bowen, 'The Elemental Composition of Human Tissues and Body Fluids', Verlag Chemie : Weinheim, New York, 1st ed., 1978.
26. H. Vanhoe, and C. Vandecasteele, Unpublished results.
27. G.E.M. Hall, J.-C. Pelchat, D.W. Boomer, and M. Powell, *J. Anal. At. Spectrom.*, 1988, **3**, 791.
28. S. Brown, Unpublished results.
29. J. Versieck, L. Vanballenberghe, A. Wittoek, G. Vermeir, and C. Vandecasteele, submitted to *Biological Trace Element Research*.

THE PREPARATION OF PLANT SAMPLES AND THEIR ANALYSIS BY ICP-MS

E. J. McCurdy

NERC ICP-MS Facility, Department of Geology,
RHBNC, Egham, Surrey TW20 0EX, UK

1 INTRODUCTION

Determination of the trace-element composition of plant materials is of interest in several applications, the most important of which include: (1) assessment of the trace-element content of foodstuffs (leaves, fruits, roots, tubers, *etc.*) for nutrition studies[1,2,3]; (2) the study of nutrient requirements and diagnosis of trace-element deficiency in plants of agricultural importance[4]; (3) investigation of the magnitude and effects of trace-element contamination of the environment[5], and (4) analysis of the tissues of indigenous plants, in biogeochemical surveying and prospecting[6].

The analysis of plant materials has, in the past, tended to concentrate on those elements which could be determined easily, using proven and commonly-available analytical techniques such as atomic absorption spectroscopy (AAS). Many studies concentrated on those elements (such as the first row transition metals, Cd, Hg and Pb) with recognised biological/environmental roles, with the result that many low-abundance heavy elements remain poorly studied. These elements (*e.g.* Y, Zr, Mo, the rare-earth elements (REEs), Th and U) typically occur at concentrations between 0.001 and 0.5 $\mu g.g^{-1}$ in plant materials[7], and are therefore below the limit of detection for analytical techniques such as AAS. Many of these low-abundance elements have important roles in geochemistry, but relatively little is known of their abundance or distribution in natural biological systems. One consequence of this lack of knowledge is that many reference materials, including the plant SRMs used in this study, are not well-characterised for these low-abundance elements. In order to verify the new dissolution technique presented here, it was therefore necessary to determine a range of elements (Ti, Cr, Mn, Fe, Cu, Zn, Rb, Sr, Ba and Pb) for which reliable reference values are available. These elements were determined in addition to the low-abundance elements of primary interest.

* *Present address*: VG Elemental Ltd, Ion Path, Road Three,
Winsford, Cheshire CW7 3BX, UK

The first commercial inductively coupled plasma mass spectrometry (ICP-MS) instruments were launched in 1983[8]. Since then, as development problems have gradually been overcome, ICP-MS has increasingly come to be recognised as a useful and versatile technique for trace-element analysis[9]. The general features of modern ICP-MS instruments have been described elsewhere, as have their analytical characteristics[10]. ICP-MS is particularly suitable for the determination of low-abundance trace-elements in plant materials, by virtue of its very low limits of detection and relative freedom from the serious interferences (caused by high concentrations of Ca, K, Mg, *etc.*), which frequently occur when analysing biological materials by optical emission spectrometry[11,12]. The most commonly used sample introduction technique for ICP-MS is solution nebulisation. Alternative techniques, notably laser ablation[13] and slurry nebulisation,[14] have been investigated for the introduction of both inorganic and organic samples into ICP-MS instruments, but solution nebulisation remains the most straightforward and easily standardised technique currently available.

Difficulties associated with the dissolution of plant materials are indicated by the range of digestion procedures which have been published[15]. The use of specialist or modified digestion flasks and carefully regulated digest conditions is frequently recommended[16]. Furthermore, many procedures rely on the use of acid mixtures containing sulphuric acid. This acid should be excluded from solutions to be analysed by ICP-MS because, in addition to causing severe polyatomic ion interferences on several elements, notably Ti, V, Cr, Zn, Ga and Ge, prolonged aspiration of dilute sulphuric acid causes degradation of the nickel sampler cones commonly used in the interface region of ICP-MS instruments.

Many published digestion procedures for plant materials result in extraction of the analytes, rather than complete dissolution of the sample, necessitating a filtration step prior to analysis[17]. Incomplete sample dissolution is acceptable in the determination of elements (such as the first row transition metals) whose behaviour during extraction has been well studied. In contrast, however, relatively little is known about the behaviour of the elements of interest in this study. Complete sample dissolution was therefore considered essential, in order to ensure, as far as possible, the preparation of samples representative of the starting material. Several workers have reported decomposition of plant tissues, using high- or low-pressure microwave, or conventional steel "bomb" digestion[3]. However, the cost of the equipment required for these procedures places them beyond the budget of many laboratories and they may be impractical for routine sample preparation. For this reason, open acid-decomposition was investigated for the preparation of samples for analysis by ICP-MS.

2 EXPERIMENTAL

An acid decomposition procedure was developed, using nitric and perchloric acid in combination with hydrogen peroxide, for the complete decomposition of plant materials. This procedure was tested on three standard reference materials (SRMs) supplied by the National Bureau of Standards (NBS), now the National Institute of Standards and Technology (NIST). The reference materials used were SRM 1572 (Citrus Leaves), 1573 (Tomato Leaves) and 1575 (Pine Needles). Reagents used in the digestion were: HNO_3 (70%) Aristar, $HClO_4$ (60%) Aristar and H_2O_2 (30%, 100 volumes) Aristar (BDH Ltd., Poole, Dorset). Deionised water (25 MΩ cm^{-3}), used for sample dilution, was produced using an Elga Spectrum system. Glassware required during the digestion was cleaned using the following procedure: 1) Soak in 5% Decon 90 in an ultrasonic bath; 2) Rinse in tap water; 3) Soak in 10% nitric acid (AnalaR) for 48 hours; 4) Rinse 3 times in deionised water.

Sample preparation procedure

Samples were prepared using the following procedure. (1) Air dry at 70 °C for 24 hours. (2) Weigh 0.5 g sample into a 50 mL Pyrex beaker and add 5 mL concentrated HNO_3. Cover with a watch-glass and leave to cold soak for approximately 7 days. (3) Add a further 5 mL concentrated HNO_3, replace watch-glass and heat at 80 °C for 4 hours, then at 180 °C for 5 to 6 hours, or until the volume of solution is reduced to about 5 mL. (4) Remove beaker from hotplate and allow to cool for one minute. Add 1 mL H_2O_2 dropwise to the solution. Care is essential at this stage as the reaction of the H_2O_2 in the hot acid is very vigorous. (5) After the reaction has died down, replace the watch-glass and return beaker to hotplate. Allow the evolution of brown NO_2 fumes to die down and then repeat the H_2O_2 addition a further four times, allowing the solution to cool for 1 minute before each addition. (6) After the final addition of H_2O_2, allow the solution, which should be colourless or pale green/yellow, to cool. (7) Add 5 mL $HClO_4$ and heat at 180 °C for 1 hour. Remove watch-glass and fume off $HClO_4$ until approximately 1 mL remains (around 1½ hours). (8) Allow solution to cool for 1 minute, then add 2 mL H_2O_2, 1 mL at a time. After each 1 mL addition of H_2O_2, return solution to hotplate until the effervescence has died down. (9) Cool and add 15 mL deionised water, then heat gently for up to 5 minutes, until the solution clears. Following dilution to 50 mL, store sample in polythene bottle prior to analysis.

Digest solutions of SRM 1572 and 1575 were clear, whilst the SRM 1573 digest had a slight precipitate, apparently of silicate material. Solutions were not filtered prior to analysis.

Sample dilution

When preparing solutions for analysis by ICP-MS, an important consideration is the concentration of total dissolved solids (TDS) present in the solution when analysed. As a general rule, this concentration should not exceed 0.1% (1000 μg.mL^{-1}) as, above this level, sampler cone blockage may occur, which can lead to severe analyte signal drift[18]. The TDS concentration of an organic sample is difficult to assess, since the organic material is composed principally of C, N, O and H. Most of the starting material mass will be lost during the digestion process, leaving only the inorganic components in the solution as analysed. A suitable dilution factor for the samples analysed in this work, was calculated on the basis of the known major element composition of the SRMs (Table 1). In order to keep the TDS concentration at approximately 0.1% or below, samples were analysed at a final concentration of 0.5 g dry weight in 50 mL (100 x dilution).

Table 1 Reference concentrations[22] (μg.g^{-1} dry weight) of major inorganic elements in plant SRMs

Element	SRM 1572	SRM 1575	SRM 1573
Ca	31500	4100	30000
K	182000	3700	44600
Mg	5800	1220	7000
S	4070	132	6200
Si	1900	810	3000
P	1300	1200	3400
Al	90	550	1200
"Total"	62860	12900	95400

Sample analysis

All samples were analysed using a PlasmaQuad PQ2 (VG Elemental, Winsford, Cheshire, UK) operated in the multi-element scanning mode. The only modification to the standard instrument was the use of a De Galan v-groove, high dissolved solids nebuliser and a single-pass water-cooled spray chamber maintained at 4 °C[19]. The use of a chilled spray chamber has been found to reduce the levels of potentially interfering oxide ions which appear in the ICP-MS spectrum[19]. The instrument was optimised on ^{115}In using the method of Gray and Williams[20], and the operating conditions used are summarised in Table 2.

Samples, including relevant digestion blanks, were analysed in batches of up to five (each sample analysed in triplicate). Calibration was by multi-element standards containing the analytes at a concentration appropriate to that expected in the samples. Three standard solutions were prepared, containing: (a) the minor elements (Ti, Cr, Mn, Fe, Cu, Zn, Rb, Sr, Ba and Pb) at 100 ng.mL^{-1}; (b) the REEs at 10 ng.mL^{-1}, and (c) the other trace elements of interest (Li, Y, Zr, Mo, Sb, Cs, Th and U) at 10 ng.mL^{-1}. The trace element standards were prepared at low concentration to reduce the likelihood of sample contamination due to analyte carry over from the standard.

Table 2 ICP-MS operating conditions

Plasma forward power	1300 W
Plasma reflected power	5 W
Coolant gas flow	14 L.min^{-1}
Auxiliary gas flow	0.5 L.min^{-1}
Nebuliser gas flow	0.75 L.min^{-1}
Sample uptake rate	1.0 mL.min^{-1}
Ion lens settings	optimised on ^{115}In.
Background counts	20
Count rates for Ce (1 μg.mL^{-1})	
M^+	1.2 M counts
MO^+:M^+	0.7 %
M^{2+}:M^+	0.15 %

All 3 standard solutions were run before and after each batch of samples, to permit assessment of analyte signal change over the analysis run. Analyte signal drift was found to be within the usual range (between 5 and 10% between standards) for samples containing this level of TDS, so calibration was based on the mean of the two bracketing standards.

3 RESULTS AND DISCUSSION

Analyte signal drift was not excessive throughout the analysis run, indicating that the TDS concentration of the samples as analysed did not cause significant matrix deposition on the components of the ICP-MS interface. This implies that the dilution factor used (100 x) is appropriate for routine ICP-MS analysis of plant samples prepared using this digestion procedure. Indeed, further samples of SRM 1572 were prepared at a dilution of 50 x (nominally 2% w/v TDS) and were analysed without causing excessive signal drift. The data from these samples have yet to be evaluated, but may indicate reduced detection limits, allowing the determination of those elements (notably some of the low-abundance heavy REEs) which were below the limit of quantitation (LOQ - see below) in some of the 1% TDS samples.

Digest blank analysis.

Although high-purity reagents were used throughout this study, some elements were found to be present at significant levels in the reagent blank (Table 3).

Table 3 Concentrations of elements detected in the digest blank.

Element	Concentration ($ng.mL^{-1}$)
B	36
Cr	10
Fe	52
Ni	25
Cu	1.3
Zn	15
Sn	17
Pb	1.5

These contaminating elements are probably derived from either the HNO_3 or $HClO_4$, since metal impurities catalyse the decomposition of H_2O_2[15]. Boron and lead are thought to be leached from the Pyrex beakers during sample digestion, so experiments were carried out to assess the use of PTFE beakers in order to avoid this contamination. The use of PTFE beakers does not appear to be viable for this particular digestion technique, as some of the plant oil residues, driven off during the decomposition, tend to adhere to the PTFE and cannot then be redissolved.

Calculation of quantitation limits

Limits of Quantitation (LOQ)[21] were calculated as 10σ (from 10 replicate analyses) of the concentration for each analyte measured in a digest blank solution. The dilution factor of 100 was taken into account and LOQ are quoted here as $ng.g^{-1}$ in the solid (Table 4). The LOQ represents the lowest concentration at which an element can, in practice, be measured quantitatively in a sample. The figures presented in Table 4 indicate the very low LOQ which are characteristic of ICP-MS, with particularly low values found for those elements above 80 amu.

Table 4 ICP-MS limits of quantitation (LOQ)

Element	Isotope (amu)	Abundance (%)	LOQ ($ng.g^{-1}$)
Li	7	92.5	68
Ti	47	7.5	164
Cr	52	83.8	340
Mn	55	100	55
Fe	56	91.7	1493
Cu	63	69.1	469
Zn	68	60	385
Rb	85	72.2	6.8
Sr	88	82.6	16
Y	89	100	18
Zr	90	51.4	41
Mo	98	24	25
Sb	121	57.3	15
Cs	133	100	5.4
Ba	137	11.3	38
La	139	99.9	7.9
Ce	140	88.5	21
Pr	141	100	6.4
Nd	146	17.3	51
Sm	147	15.1	31
Eu	151	47.8	6.4
Gd	157	15.7	27
Tb	159	100	4.3
Dy	163	24.9	29
Ho	165	100	5.6
Er	167	22.9	15
Tm	169	100	4.5
Yb	172	21.9	20
Lu	175	97.4	8.2
Pb	208	52.4	99
Th	232	100	5.5
U	238	99.3	9.4

Precision and accuracy of ICP-MS determinations

For the more abundant analytes (*e.g.* Rb, Sr, Ba and Pb), the instrumental precision for ICP-MS determinations was better than 5% RSD. With less-abundant analytes, such as Mo, Sb, the low-concentration REEs and U, precision was between 5% and 15% RSD.

All reference and ICP-MS data are presented as the mean concentration, together with the standard deviation (in brackets) where available (n = 3 for ICP-MS values). Those reference values marked with a "***" are certified by NIST[22]. All other reference values are either NIST recommended values or consensus values. These non-certified values are provided for information only, and may not be as accurate as those certified by NIST[22].

Reference and ICP-MS values for the minor elements, which were determined in order to verify the digestion procedure, are compared in Table 5. In general, ICP-MS values were in reasonable agreement with the reference concentrations. ICP-MS determinations for Ti did not agree well with the reference value for any of the SRMs, however, possibly due to inaccurate reference values. The reference values for Ti in all three SRMs are "consensus values" derived from previously published data, rather than NIST certified or recommended values. The reference concentrations for Ti in SRM 1572 and 1575 are based, in each case, on a single determination, while the reference concentration for Ti in SRM 1573 is based on three determinations (by three different techniques), covering a concentration range from 12.6 to 89 $\mu g.g^{-1}$. This particular element illustrates the problems of assessing new trace-element data by comparison with reference data, the source and quality of which may be uncertain.

ICP-MS and reference concentrations are also compared for the REEs (Table 6) and the other trace elements of interest (Table 7). Again, most ICP-MS determinations are in agreement with reference values, where reliable reference values are available. One notable exception occurs for U in SRM 1573 (Table 7), the ICP-MS value for which is significantly below the reference concentration. This may be due to the loss of U from this sample as a result of co-precipitation.

Imprecise values are a particular feature of the reference concentrations available for the REEs in all three SRMs. None of the REE reference values are certified by NIST, and most have no indication of their statistical uncertainty (see Table 6). Many of these non-certified reference values are based on only one or two determinations[22]. This makes it difficult to assess the accuracy of the ICP-MS values for the REEs. However, it is possible to assess the internal consistency of REE data sets, through the use of chondrite-normalised patterns.

Chondrite-normalised plots are an accepted method for the presentation of REE data for geological materials[eg 23] and their use has been advocated as a means of assessing the internal consistency of REE data sets obtained for such materials[24]. Chondrite-normalisation relies on the fact that, due to their similar chemical characteristics, the REEs (La, Ce, Pr, Nd, Sm, Eu, Gd, Tb, Dy, Ho, Er, Tm, Yb and Lu) tend to behave geochemically as a coherent group, with the exception of Ce and Eu which may exist in oxidation states other than the usual trivalent form[23]. The method is used to compensate for the natural pattern of REE abundance (alternate high and low concentrations) by dividing the concentration found for each REE by their average value, in chondritic meteorites[25]. A plot of the resulting values against the atomic number of the REE should result in a smooth curve, with possible anomalies for Ce and Eu, due to their variable oxidation states.

Table 5 Comparison of ICP-MS and reference minor element values ($\mu g.g^{-1}$) for plant SRMs.

Element	SRM 1572		SRM 1575		SRM 1573	
	Ref[22]	ICP-MS	Ref[22]	ICP-MS	Ref[22]	ICP-MS
Ti	22.0 (0.4)	3.4 (0.1)	13.7 (39.0)	6.5 (1.7)	56.0	37.3
Cr	0.8 (0.2)	1.2 (0.1)	2.6 (0.2)	2.7 (0.2)	4.5 (0.5)	4.9 (0.3)
Mn	23.0 (2.0)	19.8 (1.3)	675 (15)	717 (13)	238 (7)	303 (13)
Fe	90.0 (10.0)	98.1 (3.8)	200 (10)	197 (6.3)	690 (25)	761 (17)
Cu	16.5 (1.0)	16.4 (0.8)	3.0 (0.3)	2.7 (0.6)	11.0 (1.0)	12.5 (0.2)
Zn	29.0 (2.0)	32.9 (1.7)	67.0 (9.0)	66.7 (3.6)	62.0 (6.0)	75.9 (1.2)
Rb	4.8 (0.1)	5.0 (0.2)	11.7 (0.1)	12.0 (0.5)	16.5 (0.1)	18.5 (0.2)
Sr	100 (2)	108 (4)	4.8 (0.2)	4.8 (0.1)	44.9 (0.3)	51.8 (0.3)
Ba	21.0 (3.0)	19.5 (0.7)	7.2 (0.8)	8.0 (0.5)	57.0 (9.0)	64.1 (1.7)
Pb	13.3 (2.4)	11.7 (0.5)	10.8 (0.5)	10.8 (0.1)	6.3 (0.3)	6.3 (0.1)

Standard deviations (σ_n) in parentheses.

The levels of the REEs in the plant materials used in this study were generally much lower (between 2.0 and 0.04 times chondrite levels) than the levels commonly determined in geological materials[24]. Concentrations of several of the REEs (Eu, Gd, Tb, Yb and Lu) were not determined in the sample of SRM 1575 (Pine Needles), resulting in an incomplete data set for this material.

Table 6 ICP-MS and reference values ($ng.g^{-1}$) for the REEs in plant SRMs.

Element	SRM 1572		SRM 1575		SRM 1573	
	Ref[22]	ICP-MS	Ref[22]	ICP-MS	Ref[22]	ICP-MS
La	190	161 (17)	200	116 (14)	900	628 (7)
Ce	280	377 (12)	400	166 (22)	1600	1260 (15)
Pr	ND	43 (4)	<70	26 (6)	187	137 (7)
Nd	317	218 (4)	164	118 (13)	620 (70)	529 (40)
Sm	52	50 (11)	20 (2)	23* (4)	92 (16)	71 (16)
Eu	10	8 (1)	6	ND	40	11 (2)
Gd	39	43 (9)	28	ND	75	52 (5)
Tb	9	6 (2)	31	ND	9 (5)	6 (1)
Dy	43	45 (7)	ND	12* (3)	68	59 (9)
Ho	8	9 (2)	ND	5* (2)	13	9 (2)
Er	22	16 (5)	ND	15 (6)	51	25 (8)
Tm	ND	3* (1)	ND	2* (1)	ND	4* (1)
Yb	12	9* (2)	18	ND	63 (16)	21 (2)
Lu	2	1*	2	ND	9	ND

Standard deviations (σ_n) in parentheses. ND = not determined.
* = < ICP-MS LOQ.

Chondrite-normalised plots of the ICP-MS data are presented for the REEs in SRM 1572 and 1573 (Figures 1 and 2), together with plots produced from a compilation of available reference values in each of these SRMs. Lu was not determined in SRM 1573. In Figures 1 and 2 (ICP-MS data), the points marked + represent values determined at below the calculated ICP-MS LOQ. In these plots, the points representing the low-abundance, heavy REEs (Tb, Ho, Tm and Lu) are based on peak integrals of only 10 to 30 counts above background, compared with peak integrals of around 50 counts for the other (high-abundance) heavy REEs (Gd, Dy, Er and Yb). The points for these low-abundance, heavy REEs were, therefore, bracketed and omitted from the normalised plots. Considering the low levels at which the REEs were measured, the chondrite-normalised plots in Figures 1 and 2 (ICP-MS data) are exceptionally smooth, particularly in the region covering the relatively high-abundance light REEs (La to Gd), indicating excellent consistency in the ICP-MS data sets.

Table 7 Comparison of ICP-MS and reference trace element concentrations (ng.g^{-1}) for plant SRMs

Element	SRM 1572		SRM 1575		SRM 1573	
	Ref[22]	ICP-MS	Ref[22]	ICP-MS	Ref[22]	ICP-MS
Li	230 (105)	191 (30)	340	136 (38)	ND	658 (15)
Y	ND	334 (4)	ND	84 (16)	ND	295 (14)
Zr	ND	71 (8)	ND	183 (9)	ND	654 (37)
Mo	170** (90)	190 (29)	150 (50)	112 (8)	530 (90)	546 (39)
Sb	40	28 (4)	200	132 (16)	36 (7)	39 (6)
Cs	98	92 (6)	110 (10)	130 (4)	57 (8)	47 (9)
Th	ND	10 (1)	37** (3)	30 (3)	170** (30)	152 (4)
U	40 (2)	37 (5)	20** (4)	16 (3)	61** (3)	38 (5)

Standard deviations in parentheses. ND = not determined.
** = NIST certified value.

It has been pointed out that the reference values for the REE in certain geological SRMs do not form a smooth curve when plotted on chondrite-normalised diagrams, indicating inaccuracies in some of the reference values[24]. This is also the case for the plant SRMs used in this study (Figures 1 and 2), although this may not be surprising, in view of the low levels of the REE present in these SRMs.

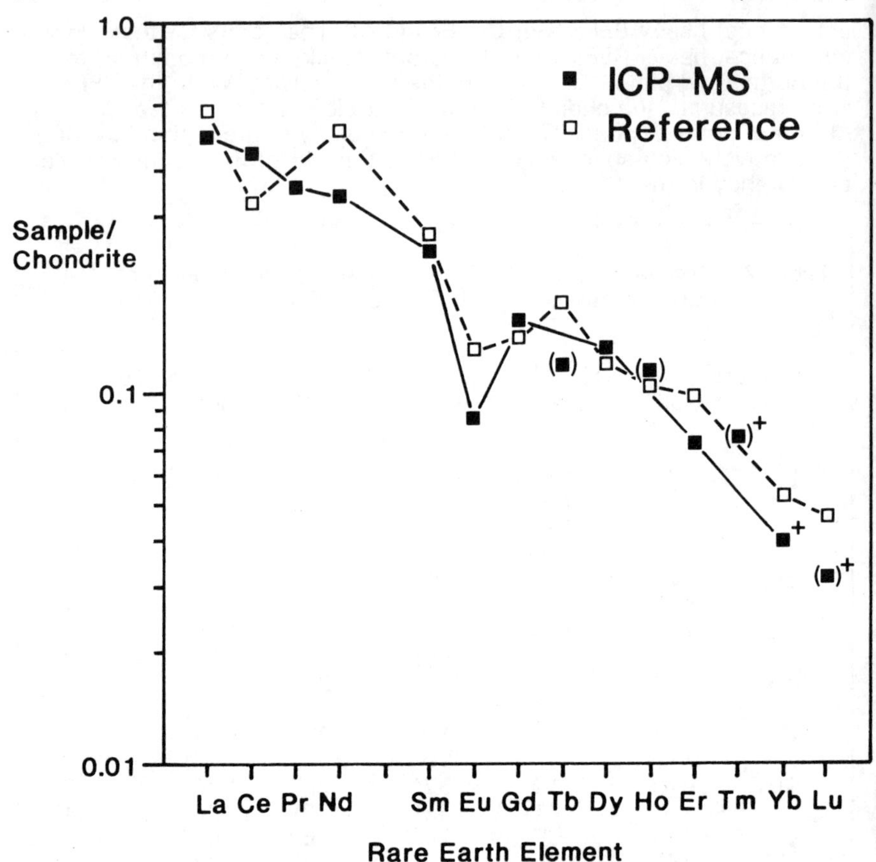

Figure 1 Chondrite-normalised plots for the rare-earth elements in SRM 1572; a comparison of ICP-MS values and reference concentrations[22]. + = value was determined at below the calculated ICP-MS LOQ. Points in brackets may be unreliable due to low count rates

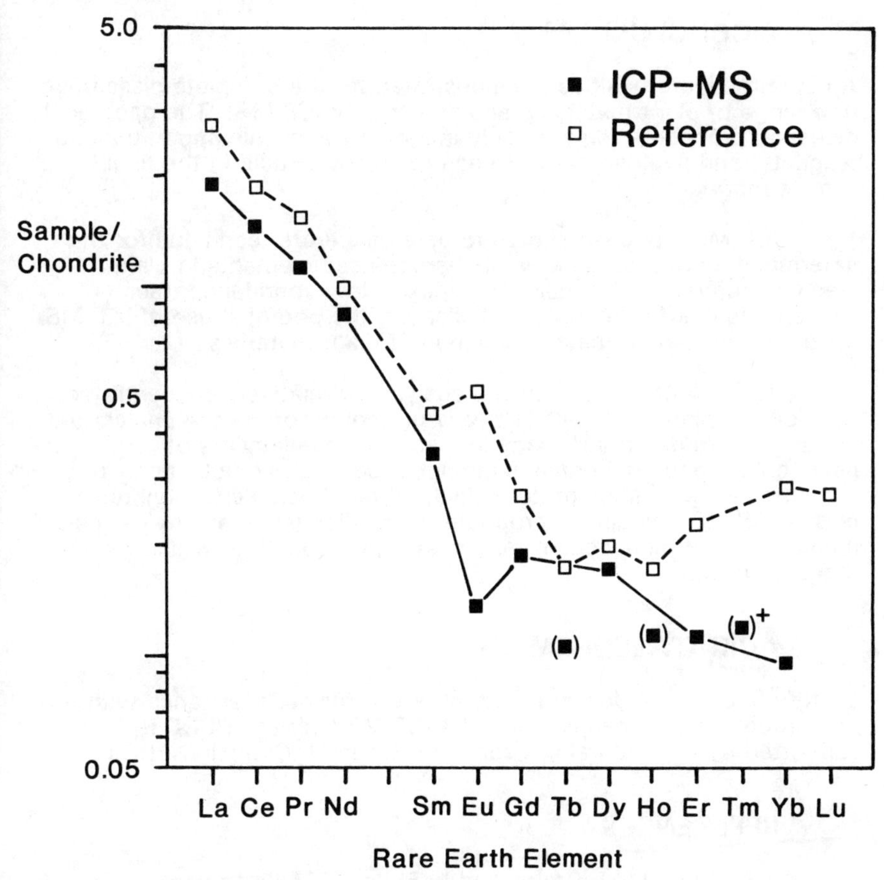

Figure 2 Chondrite-normalised plots for the rare-earth elements in SRM 1573; a comparison of ICP-MS values and reference concentrations[22]. + = value was determined at below the calculated ICP-MS LOQ. Points in brackets may be unreliable due to low count rates

4 CONCLUSIONS

A new procedure has been demonstrated for the complete dissolution of a range of plant materials, and analysis by ICP-MS. The open acid digestion procedure required only commonly available apparatus and reagents, and avoided the presence of sulphuric acid in the final sample matrix.

ICP-MS has been shown to be an accurate technique for the determination of a range of well-characterised elements in plant standard reference materials. A series of low-abundance trace elements was also determined, indicating the potential use of ICP-MS for the trace-element characterisation of plant materials.

The use of ICP-MS for the analysis of relatively concentrated solutions of plant materials (1% w/v or above), offers the analyst the opportunity to determine many low-abundance elements of geochemical and environmental importance. The combination of exceptionally low limits of detection and rapid sample throughput, makes ICP-MS uniquely appropriate for studies (such as environmental monitoring and biogeochemical surveying/prospecting) which require this type of data.

ACKNOWLEDGEMENT

Grateful thanks are due to Dr. Kym Jarvis, for her assistance with the preparation of this manuscript. The ICP-MS Unit at RHBNC is supported by the Natural Environment Research Council (NERC).

REFERENCES

1. J.R. Dean, H.M. Crews and L. Ebdon, 'Applications in Food Science', Chapter 6, in 'Applications of Inductively-Coupled Plasma Mass Spectrometry', eds. A.R. Date and A.L. Gray, Blackie, London, 1989.
2. A.G. Hugunin and R.L. Bradley, Jr., *J. Milk Food Technol.*, 1975, **38**, 354.
3. W.J. Adrian, *Analyst*, 1973, **98**, 213
4. E. Epstein, 'Mineral Nutrition of Plants:Principles and Perspectives' John Wiley and Sons Inc., New York, 1972.
5. D. Purves, 'Trace-Element Contamination of the Environment', (Revised Edition), Elsevier Science Publishers, Amsterdam, 1985.
6. R.R. Brooks, 'Geobotany and Biogeochemistry in Mineral Exploration', Harper and Row, New York, 1972.
7. H.J.M. Bowen, 'Environmental Chemistry of the Elements', Academic Press, London, 1979.
8. A.L. Gray, *J. Anal. At. Spectrom.*, 1986, **1**, 403.

9. A.R. Date and A.L. Gray, eds. 'Applications of Inductively Coupled Plasma Mass Spectrometry', Blackie, London, 1989.
10. A.L. Gray, 'The Origins, Realisation and Performance of ICP-MS Systems', Chapter 1, in 'Applications of Inductively Coupled Plasma Mass Spectrometry', eds. A.R. Date and A.L. Gray, Blackie, London, 1989.
11. S.H. Tan, and G. Horlick, *Appl. Spectrosc.*, 1986, **40**, 445.
12. A.L. Gray and J.G. Williams, *J. Anal. At. Spectrom.*, 1987, **2**, 81.
13. A.L. Gray, 'ICP-MS Analysis of Solid Samples Introduced by Laser Ablation'. Paper presented at the Kingston Conference on 'Plasma Spectrometry in the Earth Sciences: Techniques and Future Trends', Kingston Polytechnic, UK, 19-21 July, 1989.
14. J.G. Williams, A.L. Gray, P. Norman and L. Ebdon, *J. Anal. At. Spectrom.*, 1987, **2**, 469.
15. R. Bock, 'A Handbook of Decomposition Methods in Analytical Chemistry', Blackie, London, 1979.
16. R.J. Hall and P.L. Gupta, *Analyst*, 1969, **94**, 292.
17. S. Ng, M. Munroe and W. McSharry, *JAOAC*, 1974, **57**, 1260.
18. A.R. Date and K.E. Jarvis, 'Application of ICP-MS in the Earth Sciences', Chapter 2, in 'Applications of Inductively Coupled Plasma Mass Spectrometry', eds. A.R. Date and A.L. Gray, Blackie, London, 1989.
19. J.G. Williams, Ph.D. Thesis, University of Surrey, 1989.
20. A.L. Gray and J.G. Williams, *J. Anal. At. Spectrom.*, 1987, **2**, 599.
21. P.J. Potts, 'A Handbook of Silicate Rock Analysis', Blackie, London, 1987.
22. E.S. Gladney, B.T. O'Malley, I. Roelandts and T.E. Gills, 'NBS Special Publication 260-111: Standard Reference Materials: Compilation of Elemental Concentration Data for NBS Clinical, Biological, Geological and Environmental Standard Reference Materials', U.S. Department of Commerce/National Bureau of Standards.
23. K.E. Jarvis, *Chem. Geol.*, 1988, **68**, 31.
24. K.E. Jarvis and I. Jarvis, *Geostand. Newslett.*, 1988, **12**, 1.
25. N. Nakamura, *Geochim. Cosmochim. Acta*, 1974, **38**, 575.

A BASIC STUDY ON THE APPLICATION OF TETRAMETHYL-AMMONIUM HYDROXIDE (TMAH) ALKALINE DIGESTION FOR THE DETERMINATION OF SOME VOLATILE ELEMENTS BY ICP-MS

T. Cho, I. Akabane and Y. Murakami

Research Laboratory
Tama Chemicals Co. Ltd.
Tokyo, Japan

1 INTRODUCTION

When using ICP-MS or ICP-AES, the preparation of homogeneous sample solutions is required. For the preparation of the sample, an acidic digestion method has been widely applied using nitrc acid, hydrochloric acid or perchloric acid.[1] Consequently, the accurate determination of such volatile elements as B, As, Sn, Sb, Mo, Hg, I and Br have been hampered by this sample preparation method.[2,3] To overcome the problem of volatility, various methods have been used, namely, Kjeldahl flask method using a reflux condenser, oxygen combustion method, quartz tube combustion method, ion exchange method and extraction method etc.[4] These techniques are however difficult to apply widely

Alkali fusion methods have been applied to solubilize samples but are generally unsuccessful because of the lack of suitable high-purity reagents and also higher matrix effects of metallic salts (eg Na) on ICP-AES and ICP-MS. However, this company has succeeded in producing ultra-pure Tetramethylammonium hydroxide (hereafter referred to as TMAH), which has mainly been used in the semiconductor industry(U.S. patent 4,634,509).

We have applied a TMAH-alkaline digestion method to solubilize several standard reference materials, namely, bovine liver, mussel and hair, and have determined such trace elements as As, B, Ba, Cd, Co, Cr, Cu, Mn, Mo, Ni, Pb, Sn, Sr, Ti, and V on ICP-AES and GF-AAS. The values obtained agree fairly well with the certified values or reference values.[5] We considered therefore that TMAH was found to be quite suitable for alkaline digestion of biological materials.

In this study, we examined the ICP-MS operation required for the analysis of samples prepared by the TMAH digestion method. In general, on ICP-MS, nitric acid solutions are preferred in view of low viscosity and less likely formation of molecular ions. In our experiments, TMAH-alkaline digested sample solution was used as the feed solution.[6] TMAH decomposes on heating(over 130°C) and its decomposition products tend to go out of the system without any remains. There are therefore few potential matrix effects from TMAH itself. Since matrix effects of TMAH-alkaline digested sample solutions could be quite different from those of acidic digestions, the applicability of TMAH to ICP-MS has been investigated, paying particular attention to solution viscosity and molecular ion formation.

The following problems were investigated; TMAH's solubilization efficiency of biological materials, optimum concentration and its matrix effect, memory effect of TMAH-digested sample solutions, applicability of internal standard elements in TMAH alkaline medium, linearity of calibration curves, semi-quantitative response curve in TMAH alkaline medium in order to establish a set of appropriate operating conditions for ICP-MS. ICP-MS detection limits were defined in a TMAH medium.

It was found in the course of these experiments, that the washing effect of TMAH can actually be beneficial during an analytical run. After a set of operating conditions had been established, the accuracy of the technique was tested by the determination of some volatile elements in biological reference material bovine liver, mussel, hair and sargasso. Particular attention was focused on ^{127}I and ^{81}Br which can not be determined accurately by acidic digestion.

Certified values for the volatile elements are not available for the SRMs standard, therefore, values were compared with those of radioactivation analysis to judge the reliability of the proposed method.

2 REAGENT, INSTRUMENT, BIOLOGICAL SAMPLES USED

Reagent

TMAH 25% solution: TAMA CHEMICALS CO., LTD. TAMAPURE-AA-1000 grade. $(CH_3)_4NOH$, M.W. 91.15, colorless, transparent and odorless, white crystal of pentahydrate, highly soluble in water and alcohol, commercially available in 25% solution. Specific gravity; 15% d_{20}^{20} 1.011, 25% d_{20}^{20} 1.02. Viscosity at 20°C; 1% 0.923 cSt, 5% 0.998 cSt, 10% 1.174 cSt. The viscosity of the solution increases with TMAH concentration. Viscosity of TMAH is low compared to 1.01 cSt of 20°C water, making it suitable for solution nebulisation. TMAH is an organic chemical having strong alkalinity equal to NaOH and KOH; pH 12.9 for 1%, 13.7 for 5% and 14.0 for 10% solution.

As to the purity level, the TMAH of highest-purity is available with metallic impurities all below 1 $ng.mL^{-1}$ and Br and I below detection limits by ion chromatography (Table 1). TMAH is stable at room temperature but decomposes at 130-140°C into trimethylamine and dimethyl ether.

Table 1 Analytical Results of Impurities Contained in TAMAPURE-AA-1000 TMAH

$ng.mL^{-1}$

Elements	Value	Elements	Value
Sb Bi Cd[a)] Au Hg Ag Sr Th U	<0.05	B[b)] Mn Ni[a)]	<1
Al[a)] As Ba Co Cu Ga Br I Mg[a)]	<0.5	Fe[a)]	0.2
Zn 0.33 Na[a)] 0.53 Ca[a)]	0.72	Cl[c)]	<50

Values determined by ICP-MS, a) GF-AAS, b) ICP-AES using ETV, c) IC.

Ultrapure water, HNO 68%: TAMAPURE-AA-100 grade with all metallic impurities below 0.1 $ng.mL^{-1}$.
Metallic element standard solution: AAS standard solution 1000 $\mu g.mL^{-1}$, Bromine standard solution: Ion chromatography standard solution 1000 $\mu g.mL^{-1}$, Iodine standard solution:KI 13.15g (Merck, Superpur 99.5%) and $NaHCO_3$ 0.3mg are dissolved with 1 liter of ultrapure water.

Instrument

ICP-MS: PlasmaQuad PQ2, VG Elemental, Peristaltic pump: Ismatec REGLO 100 type, Nebulizer (MEINHARD type), spray chamber, elbow, torch (quartz), sampling cone, skimmer cone (nickel).

Biological Samples Used

Standard Reference Materials Used for Determination

Bovine liver NIST SRM 1755a, Mussel NIES No.6, Sargasso NIES No.9, Hair NIES No.5.

NIST; National Institute of Standards and Technology, U.S.Department of Commerce. NIES; National Institute for Environmental Studies, Japan Environment Agency.

Biological Samples Used for Dissolution Experiments. Non-fat milk powder NIST SRM 1549, Chlorella NIES No.3, Rice powder NIES No.10, Pepper bush NIES No.1. Biological samples on market: air-dried brown algae(Hizikia fusiforme), tea leaves, small fish meat, shrimp, freeze-dried cow liver and oyster.

3 EXPERIMENTS

Solubilisation of Biological Materials and Preparation of Sample Solutions using TMAH

Dissolution tests of biological samples were made by varying the TMAH concentration and heating to a range of temperatures. Closed vessel dissolution was achieved with a teflon bomb (Savillex,(R) 60mL) with a teflon vessel (PFA, 15mL) containing a sample solution. The bomb was heated at 130°C in an oven under medium pressure (up to 100 $kg.cm^2$). Sample solutions were also prepared in open teflon vessels, which were left under room temperature, or heated at 60-65°C and 70-75°C on a water bath, 80-85°C and 90°C in an oven. Solubility of each condition was examined. The heating time applied was, respectively, 10, 20, 30, 50 min., 1, 2, 3, 4, 5, 6, 7, 8, 24hrs., 2 days (room temp.). The details of these experimental results are given elsewhere.[6]

Comparison between Acidic Digestion and TMAH Digestion

The following experiments were conducted to examine to what degree the volatile elements are volatilized by acidic digestion, in comparison with TMAH digestion (Figure 1). On hair, 3 kinds of solutions each containing 0.1g of the sample were prepared as follows: The sample was added with 5 mL of 5% TMAH in a teflon vessel, after being screw-capped, heated at 80°C for 5 hrs and finally made up to 100 mL with 1% TMAH solution(a). The second sample was added with 20 mL of 68% HNO_3-20% HCl (1:3 vol) mixed solution and heated to dryness until the acid vapor was no longer evolved, and then made to 100 mL with 1% HNO_3 solution(b). The third sample was added with 20 mL of 68% HNO_3 and 1 mL of 20% $HClO_4$, and heated to dryness until white fumes of $HClO_4$ were no longer generated. It was then made up to 100mL volume

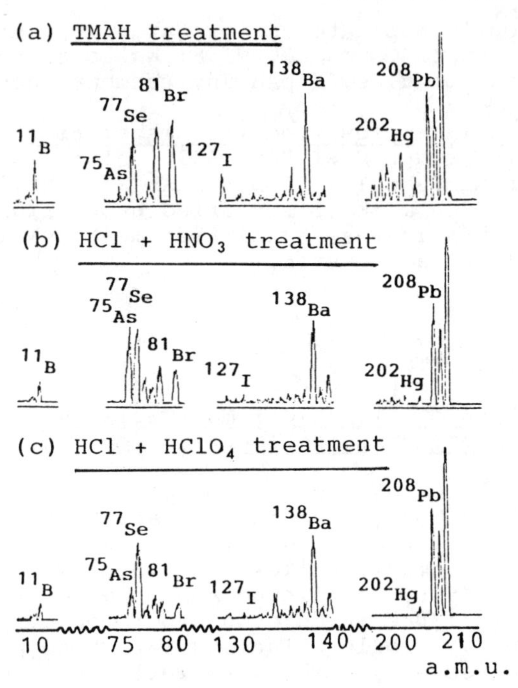

Figure 1 Comparison spectra of ICP-MS of Hair obtained by TMAH(a) and acid treatment(b,c)

with 1% HNO_3 solution(c). Each solution was analysed at a concentration of 1 $mg.mL^{-1}$. Comparing the spectra obtained from these solutions, it was observed that for elements such as B, Br, I, Hg, Pb, no losses occured in solution a). In solutions b) and c), however, more than 50% of each element appeared to have been volatilized. The use of an alkaline TMAH-digested solution is effective for the determination of volatile Br and I in SRM bovine liver, hair, mussel and sargasso by ICP-MS.

Preparation of Sample Solution

200-500 mg of the biological sample and 5 mL of 10-25% TMAH were put into a teflon vessel. The screw-capped vessel was then heated at 65-85°C for 1-2 hrs. After cooling, the sample was made up to a volume of 250 mL. In same cases, residues remained in the digested solution, these may be filtered off by alkali-proofed filter paper thoroughly washed by 5% HNO_3 and 5% TMAH before use. All preparation was carried out in a clean room of Class 1000. An outline of this procedure is shown in Figure 2. However, for small samples like biopsies, it would be necessary to use high pressure heating by microwave to effect complete dissolution.

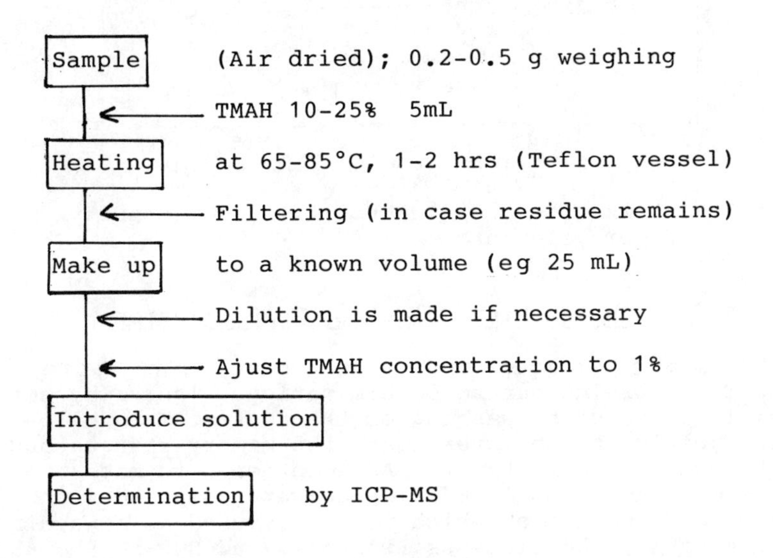

Figure 2 Procedure of sample dissolution by TMAH

TMAH Concentration and Matrix Effects

The matrix effect from TMAH, when used for biological sample digestion, was investigated. For the determination of the target elements, 100 $ng.mL^{-1}$ each of Br and I, and Tl and Bi as an internal standard were used. The following experiments were made; 1% HNO_3 (which is normally used with ICP-MS), ultrapure water and TMAH at 0.1, 0.5, 1, 2.5, 5, 10% solutions were analysed by ICP-MS at uptake rate of 0.7 $mL.min^{-1}$ and analysed on the ion intensity (cps/peak area) of ^{81}Br, ^{127}I, ^{205}Tl and ^{209}Bi. The relative intensity of each solution to that of ultrapure water was taken for each element and solution and shown in Figure 3. The intensity of TMAH can be seen to decrease slightly at concentrations over 5%.

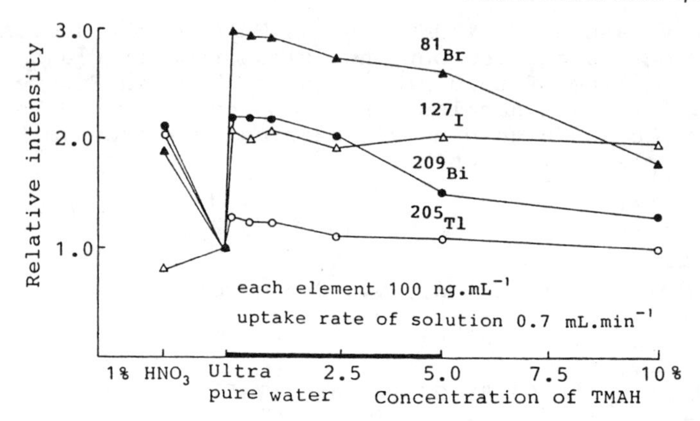

Figure 3 Matrix effect of TMAH on the ion intensity of some elements

Memory Effect and Washing Effect of TMAH Solution

The memory effects of certain elements might cause difficulties of inaccurate determinations. They may not be easily improved by washing of the system. The following 2 experiments to investigate the memory effects and also the washing effects of TMAH solution were made. The results obtained from TMAH solution were compared with those of HNO_3 solution, which is widely used as a washing agent. Firstly, the elements Br and I, and Hg and Pb (which tend to cause memory effects) were considered. 1% HNO_3 solution containing 10 $ng.mL^{-1}$ each of these elements was run for 10 min, and then, to clean the system, 1% HNO_3 solution was run for 10 min followed by 5% TMAH solution for 10 min (Figure 4).

The washing effect was examined at the following scanning mode conditions; At peak jump mode, Number of Channels 2048, Dwell time 320 μs, Point per peak 5, Dac step between point 5, Number of peak jump sweeps 50. Each measurement was repeated 20 times per 30 sec, that is, for 10min, and the average ion intensity (cps/peak area) and standard deviation were calculated from these results.

Secondly, the elements Br, I, and Tl were considered which were found to be difficult to remove by HNO solution. First, 6.8% HNO_3 solution containing 100 $ng.mL^{-1}$ each of Br, I and Tl was run for 5 min, then, 6.8% HNO_3 solution for 30 min (6.8% concentration was used instead of 5% because its original reagent is of 68%: TAMAPURE-AA-100). And then, 5% TMAH solution was run.

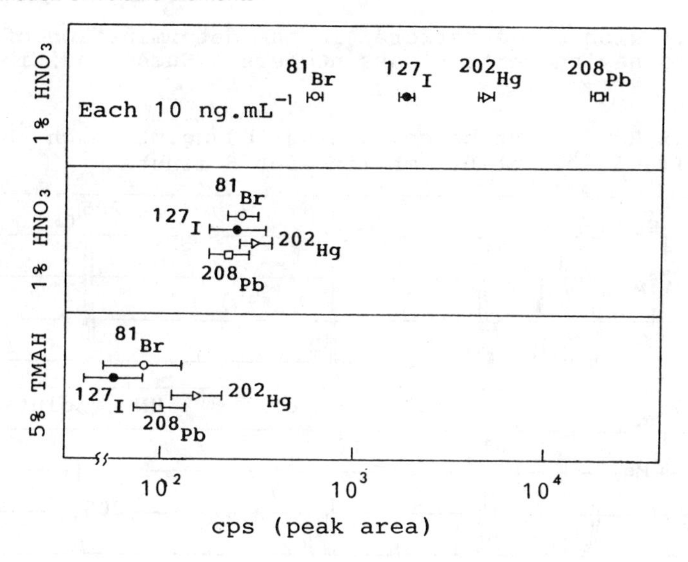

Introduced at 0.74 $mL.min^{-1}$ for 10min, respectively

Figure 4 Comparison of the washing effects of 1% HNO_3 and 5% TMAH solution for Br, I, Hg, Pb, 10 $ng.mL^{-1}$ each.

Spectra were collected each time the solutions were run (Figure 5). The intensity of ^{81}Br and ^{127}I both gradually decrease after introducing 20 mL of 6.8% HNO_3 solution (spectrum b), but sharply decline with each introduction of 5% TMAH solution (spectra c, d, e). This tendency also is observed on ^{205}Tl. This suggests that washing effect of TMAH is excellent. When TMAH solution was used in place of HNO solution as the washing solution, pulse signals around the mass range of ^{81}Br and ^{127}I decreased, gradually reaching background level. It may be said that the bigger the amplitude, the greater the washing effect.

Operating Conditions of ICP-MS

Experiments were made to determine the optimum operating conditions of the instrument where the highest intensity with stable pulse signal can be obtained, whilst the formation of molecular ions and divalent ions is lowest, which may interfere in the determination of the elements having certain mass numbers. Such conditions

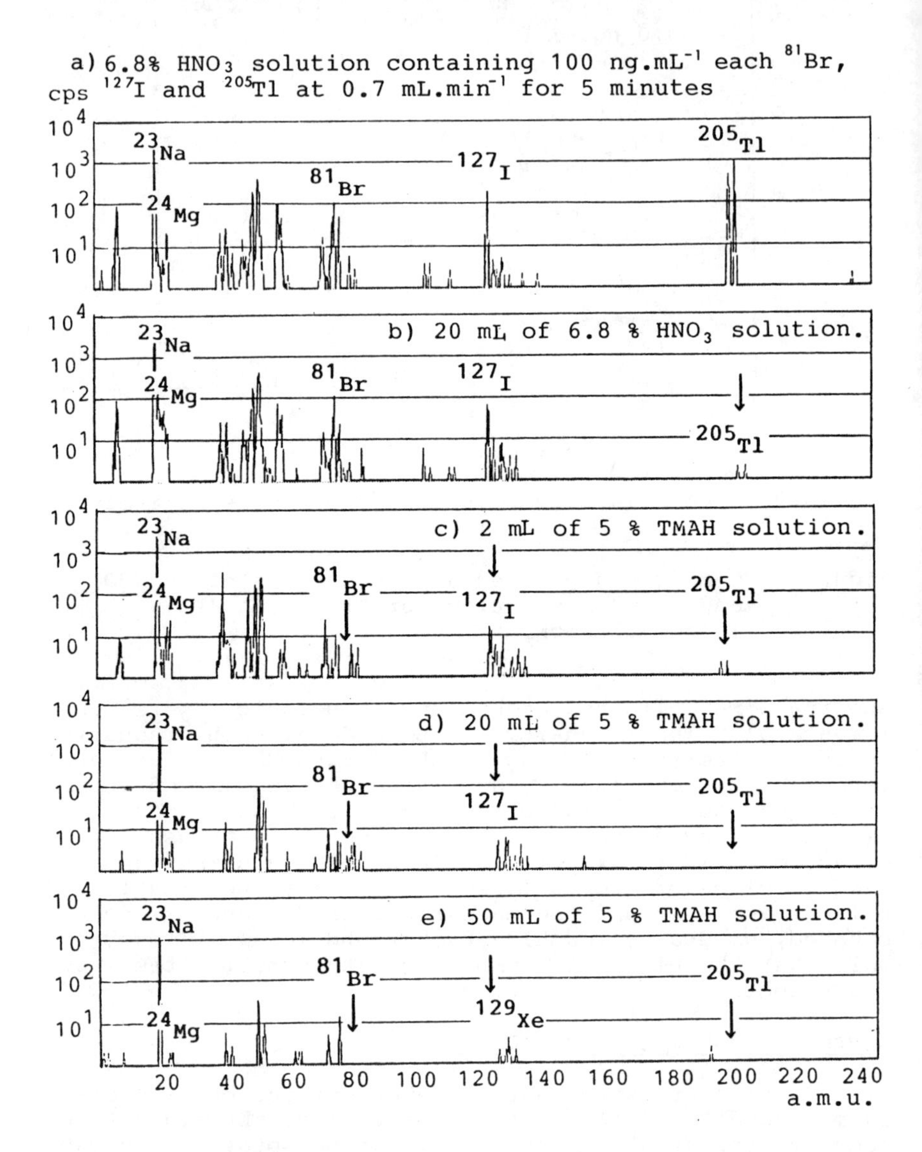

Figure 5 Decrease of peak height by washing with 6.8% HNO_3 solution (b) and increased volume of 5% TMAH solution (c, d, e).

were studied by changing the following 3 basic conditions while monitoring $^{138}Ba^{2+}$(69 amu) as a divalent ion and $^{138}Ba^{16}O^{+}$(154 amu) as an oxide ion. (a) Ar gas flow rate for the nebulizer (b) uptake rate of feed solution (c) temperature of cooling water for the spray chamber. 1% TMAH multi-element solution containing 100 ng.mL^{-1} each of B, V, Co, As, Se, Br, Sr, Y, Mo, Nb, Sn, Sb, I, Cs, La, Pr, Hg and U was used to obtain their respective ion intensity (cps/peak area).

<u>Optimum Ar Gas Flow Rate for Nebulizer.</u> The following conditions were considered: temperature of cooling water for the spray chamber (-5°C), sampling depth (distance from the tip of sampling cone to this side of load coil) 10mm, uptake rate of feed solution 0.7 mL.min^{-1}. While monitoring the formation of $^{138}Ba^{2+}$, and $^{138}Ba^{16}O^{+}$, the best condition required to obtain the highest ion intensity with stable pulse signal on each element were determined by changing the Ar gas flow rate by 0.05 L.min^{-1} within the range of 0.5-0.8 L.min^{-1}. As a result,it was observed that the ion intensity of each element showed a straight-line rise in the range of 0.5-0.65 L.min^{-} and was stable in the range of 0.65-0.75 L.min^{-1}, but declined slightly beyond 0.75 L.min^{-1}.

For $^{138}Ba^{16}O^{+}$, the ion intensity was of the background level up to 0.60 L.min^{-1}, increasing in the range of 0.63-0.73 L.min^{-1} and went up in a straight line beyond 0.74 L.min^{-1}. The ion intensity of divalent $^{138}Ba^{2+}$is lower than $^{138}Ba^{+}$ by 2 orders of magnitude, but showed the same tendency with increase in the flow rate.

It was concluded that the flow rate of 0.70-0.73 L.min^{-1} was most suitable, where higher ion intensity with stable pulse signal is found, and the formation of divalent ions and oxide ions is least likely.

<u>Uptake Rate of Feed Solution.</u> An experiment was made using the following conditions; Ar flow rate 0.730 L.min^{-1}, sampling depth 10mm, temperature of cooling water for the spray chamber -5°C.

The optimum conditions were determined by changing the uptake rate of feed solution to the nebulizer by 0.1 mL.min within the range of 0.5-1.1 mL.min^{-1}. The corresponding ion intensity of the elements increased in a straight line from 0.4 to 0.7 mL.min^{-1}, with the remainder almost at the same level at 0.7-0.8 mL.min , but decreased by about 1% each from step from 0.8 to 1.1 mL.min^{-1}. The formation of oxide ($^{138}Ba^{16}O^{+}$) was the highest 400 cps at 0.9 mL.min^{-1}, 150 cps at 0.8 mL.min , 60 cps at 0.7 mL.min^{-1} and 30 cps in the range of 0.4-0.6 mL.min^{-1}. As for $^{138}Ba^{2+}$, about 100 cps was produced at over 0.8 mL.min^{-1}, but at

0.7 mL.min^{-1} it was not measured.

Temperature of Cooling Water for the Spray Chamber.

The conditions of the experiment were: Ar gas flow rate to the nebulizer 0.730 mL.min^{-1}, uptake rate of feed solution 0.70 mL.min^{-1}, sampling depth 10mm. Ion intensity level of the elements and the formation of $^{138}Ba^{2+}$ and $^{138}Ba^{16}O^{+}$ were checked by changing the temperature of the cooling water for the spray chamber by 5°C within the range of -5 to 20°C. It was found that, at below 5°C, the formation of $^{138}Ba^{2+}$ and $^{138}Ba^{16}O^{+}$ was 50-100 cps and less than 20 cps, respectively. Consequently, no interference occurred in the analyses of the elements having mass 69 amu and 154 amu. ^{11}B, ^{51}Co and ^{202}Hg showed the highest ion intensity in the range of 5-15°C, whilst, for other elements the range was 0-5°C.

As a result of this, (and to account for the distance from the circulation equipment of the cooling water to the spray chamber being 2 m), the temperature of cooling water was set at 0°C so that the temperature of the water inside the chamber always remained below 5°C. With all the experimental data obtained, the operating conditions were fixed as shown in Table 2. These operating conditions were then applied to the following experiments.

Table 2 Typical Operating Conditions on VG PlasmaQuad PQ2

A Plasma	
Power	1.35kw
Nebulizer gas	0.73L.min^{-1}
Auxiliary gas	0.5 L.min^{-1}
Plasma gas	13 L.min^{-1}
Uptake rate of feed solution	0.7mL.min^{-1}
Sampling depth above load coil	10mm
Temperature of cooling water for spray chamber	0°C
B Mass spectrometer operating conditions	
Vacuum stage 1	2.3mbar
Vacuum stage 2	0.0X10^{-4} mbar
Vacuum stage 3	2.0X10^{-6} mbar
C Scan conditions	
Mass range	10.00-240.78a.m.u.
Number of scans	100
Dwell time per channel	160μs
Number of channels	4096

Selection of Internal Standard Elements

When ICP-MS is used for semi-quantitative analysis in accordance with the manufacturer's instruction, the use of internal standard elements are recommended. They are used to monitor and correct for fluctuations of matrix element composition, instrumental drift, nebulizer and sampler orifice blockage and aerosol transport effects. Each biological sample has a different organic composition depending on the species and different composition of major elements contained in biological samples. This necessitates the use of internal standard elements to correct for matrix effects.

The following experiments were conducted to select the most suitable internal standard elements, from In, Cs, Tl and Bi, for the determination of Br and I (Table 3).

1) For each of these elements, a calibration curve was established for the range of 0.1-1000 ng.mL^{-1} of Br and I by using 1% TMAH solution.

Table 3 The Occurrence of Elements Used for Internal Standard and their Ionization Potential

Element	Earth Crust µg.g	Soil µg.g	Sea water µg.g	Ionization Potential
Cs	3	4	3 X 10^{-4}	3.89 eV
In	0.1	1	4 X 10^{-6}	5.79 eV
Tl	0.5	0.2	1 X 10^{-5}	6.11 eV
Bi	0.2	0.2	2 X 10^{-5}	7.29 eV
Ar	---	---	---	15.76 eV

2) A TMAH-alkaline digested sample solution of sargasso was made up to 1% TMAH solution at a sample concentration of 1 mg.mL^{-1}. This was added with 0.1-1000 ng.mL^{-1} of Br and I.

Four calibration curves were calculated out on each internal standard element.

3) 100 ng.mL^{-1} of the internal standard elements was used to obtain a sufficient ion intensity level when added to the respective feed solution of 1) and 2).

As shown in Figure 6, Tl and Bi exhibited similar chemical behaviours to Br and I in 1% TMAH calibration curve solution of 1), of which the relative intensity to concentration showed good linearity, in the range 0.1 -1000 ng.mL^{-1}, with a correlation coefficient at over 0.995.

In the case of the TMAH-digested sample solution of sargasso of 2), which was selected as being representative of biological samples that are expected to contain large amounts of Br and I as marine products, good linearity with over 0.990 correlation coefficient was indicated. However, the calibration curve of Tl and Bi in 1% TMAH solution was different from that of 1% TMAH sample solution of 2), therefore, these two internal standard elements aren't suitable for the determination of Br and I. However, since In and Cs showed the same inclinations in both solutions, these elements were selected as the internal standard elements for Br and I. In the case of 1% TMAH sample solution added with 100 $ng.mL^{-1}$ each of In, Tl and Bi, the intensity level of both Tl and Bi sharply declined, suggesting some alteration in the solubilization state. Hence, it was decided to use only one Cs or In as the internal standard element for each sample feed solution.

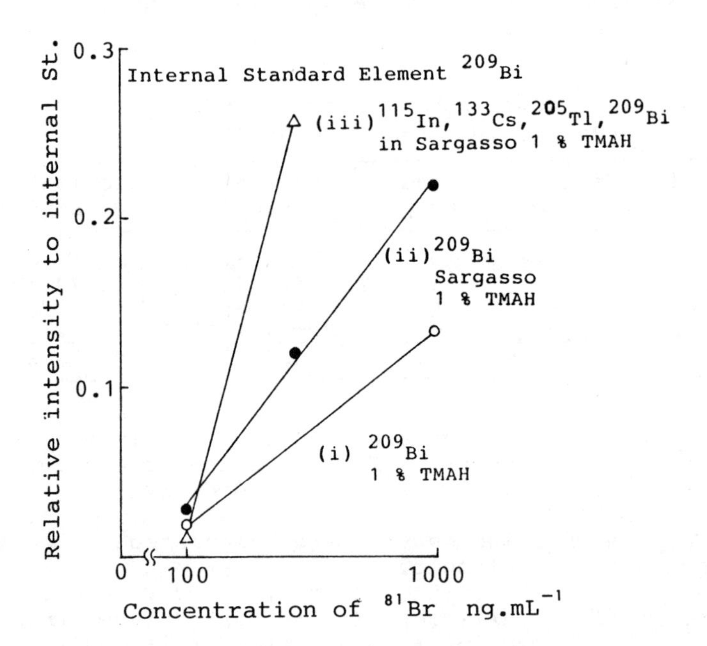

△ Sargasso 1 $mg.mL^{-1}$ 1% TMAH solution containing of In, Cs, Tl and Bi (each 100 $ng.mL^{-1}$)
● Sargasso 1 $mg.mL^{-1}$ 1% TMAH solution
○ 1% TMAH solution

Figure 6 Calibration curve of Br as the internal standard Bi

Reproducibility and Stability of TMAH Solution.

TMAH-alkaline digested sample solutions of mussel were prepared at 1% TMAH solution with a sample concent-

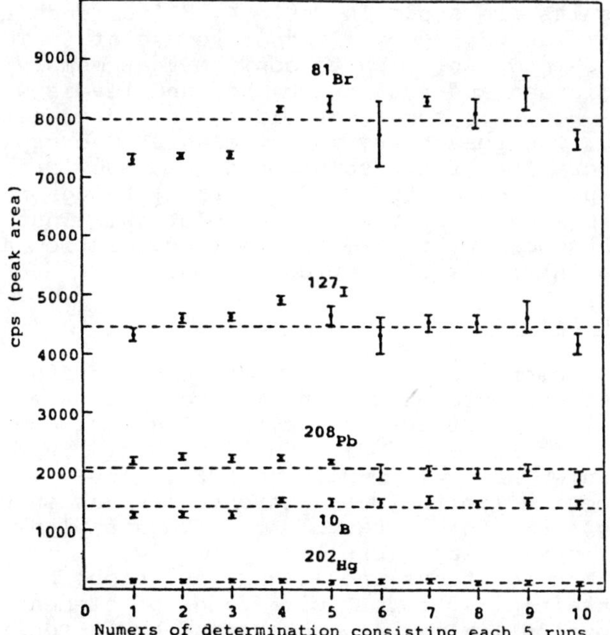

Dotted line show average of 10 determination, bar of S.D. calculated from results of 5 runs

Figure 7 Reproducibility of 4 elements determined in 1% TMAH-digested mussel solution.

Table 4 The Elapse of Time after Preparation of the Feed Solution

Element	After half day		After 3 days	
	ng.mL^{-1}	CV %	ng.mL^{-1}	CV %
^{81}Br	33.67	1.60	41.04	5.90
^{127}I	30.09	1.39	34.05	3.34
^{120}Sn	3.62	0.02	3.73	0.42
^{10}B	26.12	6.28	38.45	6.04
^{63}Cu	266.09	3.84	333.47	8.71
^{202}Hg	4.89	0.39	0.12	0.01

ration of 1 $mg.mL^{-1}$. Immediately after its preparation, the ion intensity was measured in a sequence of 5 runs and then analyzed 10 times (Figure 7). The determination was made twice, on this TMAH-alkaline digested sample solution, at 12 hrs and 3 days after sample preparation. These results are shown in Table 4. After each analysis, the system was washed by alternative use of 5% TMAH solution and 5% HNO_3 solution to confirm that memory effects were eliminated and that the background levels were recovered. However, sometimes widely scattered data were observed. The reason why is not clear. One explanation is deterioration of the solution due to the lapse of time after preparation of the feed solution, though no perceptible change was observed in the solution. Thus these experiments were performed on the feed solution at different time intervals after preparation.

Detection Limits

Measurements of ion intensity were carried out on 1% TMAH solution containing 10 $ng.mL^{-1}$ each of 18 elements to determine the detection limits. These elements were ^{11}B, ^{51}V, ^{52}Cr, ^{55}Mn, ^{59}Co, ^{60}Ni, ^{63}Cu, ^{66}Zn, ^{75}As, ^{77}Se, ^{81}Br, ^{98}Mo, ^{111}Cd, ^{120}Sn, ^{121}Sb, ^{127}I, ^{202}Hg, and ^{208}Pb.

Detection limits (D.L.) were defined as shown by the following equation: 10 $ng.mL^{-1}$ X (3 S.D. of blank)/(element solution, cps - blank, cps)

1% TMAH solutions and 1% HNO_3 solutions were prepared containing 10 $ng.mL^{-1}$ of a range of elements.

Detection limits were determined by 10 replicate analyses of a blank solution and of each solution (Table 5)

Semi-Quantitative Analysis

First, 1% TMAH multi-element solution containing 50 $ng.mL^{-1}$ each of ^{9}Be, ^{27}Al, ^{89}Y, ^{115}In, ^{139}La, ^{209}Bi and ^{238}U was prepared and measurement was carried out on scan mode to determine the instrumental mass response curve (Figure 8); Number of Channels 2048, Number of Scan Sweep 100, Dwell time 320 μs, Points per peak 5, Dac Step between points 5, Number of peak jump sweeps 20.

Secondly, 1% TMAH feed solution of an unknown biological sample was made to contain a known amount of In and analyzed using this scan mode. As a result, semi-quantitative values at the $ng.mL^{-1}$ level were found as shown in Table 6.

The semi-quantitative values given by this curve on the sample solution are considered to be accurate to within 30% of the "correct" result and can provide a rough check on quantitation for unknowns. However, the matrix effects of increased amounts of sample should be considered for use of TMAH solution.

Table 5 Comparison of Detection Limits Between 1% TMAH and 1% HNO_3

Element	1% TMAH Solution $10 ng.mL^{-1}$		1% HNO_3 Solution $10 ng.mL^{-1}$	
	Ion Intensity cps	DL* $ng.mL^{-1}$	Ion Intensity cps	DL* $ng.mL^{-1}$
^{11}B	11145±335.25	5.9	6361± 71.96	0.34
^{51}V	24718±344.43	0.17	27599± 99.11	0.016
^{52}Cr	59362±875.55	0.84	25760± 97.04	0.062
^{53}Cr	3401± 62.47	0.24	3262± 17.96	0.12
^{55}Mn	31417±399.00	0.008	37363±155.10	0.043
^{59}Co	23828±287.06	0.014	28763±130.72	0.032
^{60}Ni	5670±142.76	0.24	6722±130.72	0.018
^{63}Cu	12607±154.59	0.23	14967±177.53	0.33
^{66}Zn	5550± 66.81	0.027	6362±287.90	0.006
^{75}As	6478± 60.23	0.089	5781± 16.62	0.072
^{77}Se	664± 19.42	0.21	451± 5.12	0.42
^{81}Br	1105± 24.99	0.11	596± 31.84	3.9
^{98}Mo	9283± 90.74	0.010	10288± 69.00	0.05
^{111}Cd	4228± 65.27	0.008	4847± 47.35	0.013
^{120}Sn	16889±116.72	0.005	18101±108.83	0.007
^{121}Sb	18892± 98.53	0.024	19854±148.28	0.001
^{127}I	10442±108.71	0.016	897±886.60	0.32
^{202}Hg	25115± 14.38	0.001	8868±160.25	0.006
^{208}Pb	32990±294.07	0.005	34303±202.64	0.002

* : Detection Limits 3σ

Correlation of Sample Concentration and Ion Intensity

It might arise that the sample concentration of a target element is desired to be increased to raise the ion intensity level when it is very low. A 1% TMAH solution of sargasso was prepared, (containing high concentrations of Br and I), and concentrations of Br and I were determined in solutions containing 0.1-4 $mg.mL^{-1}$ of sample. In the case of ^{127}I, good linearity was obtained, with a correlation coefficient of over 0.98, although a slight change in the intensity was observed at around 1 $mg.mL^{-1}$ sample (Figure 9).
Therefore, it was revealed that there is no effect from sample concentration in this range. However, when the sample concentration was increased, contamination and blockage of the sampling cone and skimmer cone occurred. So, a sample concentration of 1 $mg.mL^{-1}$ was used for fur-

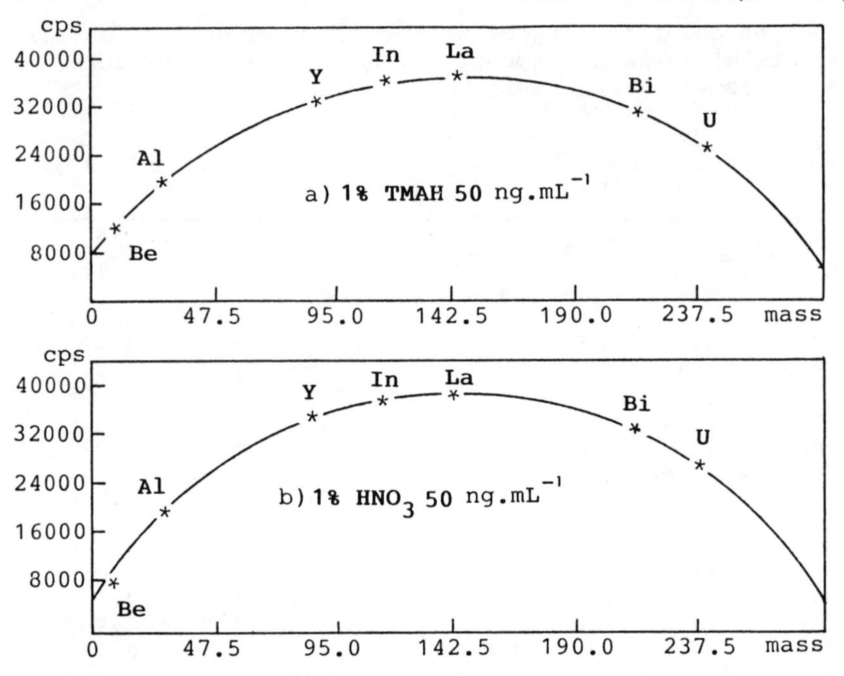

Figure 8 Instrumental mass response curve

Table 6 Example of Semi-quantitative Results obtained from the Curve a) for Some Elements in TMAH Digested Sample of Hair (NIES NO.5)

Element	Concentration $\mu g.g^{-1}$		Element	Concentration $\mu g.g^{-1}$	
Ti	30.9	(22)	Ga	0.19	
Cr	9.3	(1.4)	Br	3.8	(90)[a)]
Mn	4.2	(5)	Sr	1.69	(2.3)
Ni	1.5	(1.8)	Cd	0.35	(0.2)
Cu	15.3	(16)	I	0.068	(0.62)[b)]
Zn	274	(169)	Pb	7.6	(6)

Figures in parentheses are Certified Values.
a) Reference Values. b) Values determined by INAA

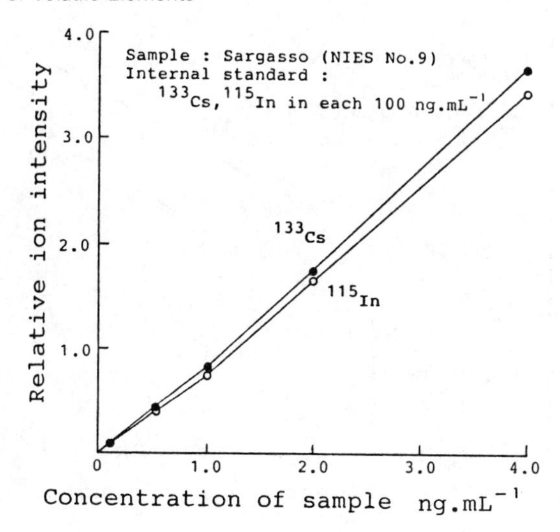

Figure 9 Effect of sample concentration on relative ion intensity of ^{127}I to that of standard ^{133}Cs or ^{115}In

ther analyses. ^{81}Br also showed the same tendency at 1 $mg.mL^{-1}$.

Linearity of Calibration Curve

To find the optimum concentration range of Br and I, where a stable pulse signal with the least formation of oxide ions or divalent ions and good linearity can be shown, 1% TMAH solution containing Br and I in the range of 0.1-1000 $ng.mL^{-1}$ was analyzed (Figure 10). Both elements gave good linearity with over 0.998 correlation coefficient.

Ion intensity of ^{81}Br is, in equal concentration, half of that of ^{127}I and, therefore, area cps are given only below 100 in the range of 0.1-1.0 $ng.mL^{-1}$. At this level the signal is unstable and precision[7] is poor. Consequently, the coefficient of variation(cv %) becomes large compared with that of ^{127}I. From the above, the concentration range of calibration curves of Br and I were decided to be 1-50 $ng.mL^{-1}$, where stable pulse signals can be shown and also memory effects do not take place.

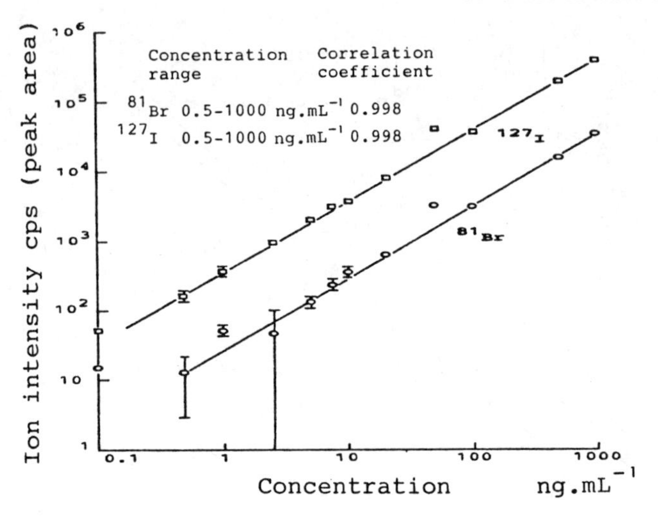

Figure 10 Calibration curves of ^{81}Br and ^{127}I

Quantitative Analysis

Since the concentration of each target element can be estimated by semi-quantitative analysis, sample feed solutions were diluted in order to contain each element in a range of several to tens of ng.mL^{-1}, where memory effects do not occur. Sample feed solutions were made to be 1% TMAH solution. By using standard solutions which were prepared to be also 1% TMAH, calibration curves for the concentration range of 1-50 ng.mL^{-1} were calculated. Both calibration curves of standard solution and of feed solution were respectively prepared to contain 100 ng.mL^{-1} each of In and Cs as the internal standard. When changing from calibration curve solution to sample feed solution, or from a sample feed solution to the other sample feed solution, this sequential washing procedure was made; ultrapure water - 5% HNO_3 solution - ultrapure water - 5% TMAH solution - ultrapure water.

After this washing, when the memory of a target element still was not returned to the background level, this sequential washing procedure was conducted again.

Radioactivation Analysis

Accurately weighed 30-400 mg of a sample dried at 80-100°C was put into a polyethylene bag and double wrapped. The sample was irradiated with thermal neutron flux, $3.2X10^{12}$ $n.cm^{-2}.s^{-1}$. for 5 hrs in a pneumatic tube or central hole, and cooled for 5-7 days for ^{82}Br and 20-40 days for ^{128}I. Then, replacing the outer polyethylene bag with a new one, the sample was again irradiated and analyzed for gamma ray. The irradiation time was 2-5 hrs and 8-40 hrs, respectively. Measurement was performed by a Ge(Li) detector (Canberra Co.) and 4096 Channel pulse height analyzer. Photo peaks used for the determination are 616.9 KeV for ^{82}Br and 442.9 KeV for ^{128}I.

These results were used to compare with those obtained by ICP-MS for TMAH digested biological sample.

4 RESULTS AND DISCUSSIONS

TMAH-Alkaline Digestion of Biological Samples and Preparation of Sample Solutions

Colorings of TMAH-alkaline digested solutions of biological samples are usually pale yellow to mud yellow (mainly due to degradation products of protein). Actual application of this method to ICP-AES and GF-AAS already provided satisfactory results.[5]

Solubilisation of biological samples is normally made by an acidic digestion method, which always carries the problem that some elements tend to volatilize. For these 18 volatile elements such as B, Br, Mo, Cd, Sn, Sb, I, Hg, Pb, etc., the following experiments were made to study the applicability of this method to ICP-MS by using standard solutions of these elements. As mentioned in the section at the Solubilisation and Preparation of samples (3 EXPERIMENTS), 14 kinds of biological samples were examined by changing the temperature, pressure and time, for which sample preparation methods are shown in Figure 2.

Advantages of TMAH and its Washing Effect

TMAH decomposes at over 130°C and is strongly alkaline. As observed from Figure 3, 1% TMAH concentration is suitable for this purpose. The formation of molecular ions of C, N and others might be a caution with TMAH application, but actually such formation did not occur.

As shown in Figure 8, the instrumental mass response curve of 1% TMAH solution is entirely the same as that of 1% HNO_3 solution that is in common use. In other words, there was no bad effect with the use of TMAH. There was also no matrix effect from TMAH (Figure 3). But, when TMAH concentration is over 5%, it leads to the decrease of ion intensity, therefore, concentration of TMAH feed solution to nebulizer was made to be below 1% (uptake rate at 0.7 $mL.min^{-1}$).

It was also recognized that there is no memory effects from TMAH itself and, on the contrary, its washing effect of other elements is quite strong. It was proven that the memories of the elements, which seldom return to background levels by even long (sometimes more than half day) washing of commonly-used ultrapure water and HNO_3 solution, can be easily removed. The results of our experiments made on ^{202}Hg, ^{208}Pb and ^{205}Tl, memories of which tend to remain, are shown (Figures 4 and 5).

Consequently, it is proposed in this report that this sequential washing procedure of ultrapure water - 5% HNO_3 solution - ultrapure water - 5% TMAH solution - ultrapure water is quite effective for eliminating memory effect with ICP-MS. TMAH was also found to be effective for removing Na, Mg and other elements which may be liable to produce contamination problems of the instruments. If TMAH is used to wash the containers used in analyses, accurate determination of even $pg.mL^{-1}$ level will be made possible.

Once there remained memories in the introduction system, torch, sampling cone and skimmer cone, they had to be all replaced and washed thoroughly for re-use. In Figure 5, the spectra measured at each time of feeding the solutions are given in a chronological order: 6.8% HNO_3 solution containing 100 $ng.mL^{-1}$ each of Br, I and Tl(a), 20 mL of 6.8% HNO_3 solution(b), 2, 20 and 50 mL of 5% TMAH solution(c, d, e). It is clearly shown that memory, which could not be removed by washing with HNO_3 solution, have been gradually eliminated with an increased volume of TMAH solution. It considered that memories are quickly washed off the components of the introduction system as teflon tube, nebulizer, chamber, elbow, torch, etc. with the feeding of TMAH solution.

Operating Conditions and Internal Standard Elements

Poor analytical precision can be attributed to changes in operating conditions or, it is sometimes thought, to the chemistry itself of the element. As stated above (3 EXPERIMENTS), the optimum operating conditions are the ones that can provide high sensitivity and stable pulse signals, and naturally do not produce molecular ions or divalent ions . From this point of view, the following three parameters were closely examined; 1) Ar gas flow rate for the nebulizer, 2) uptake rate of feed solution, 3)temperature of cooling water for the spray chamber. The experiments thereafter were all conducted under these conditions (Table 2). However, it is recommended by the manufacturer's instruction to use internal standard elements. In, Cs, Tl and Bi were selected as the internal standard element since they are not usually found in biological and environmental samples. Two kinds of calibration curves, one for each element and the other for a sample solution of biological materials containing the internal standard element, were compared. The suitable internal elements are considered to be the ones that can exhibit good linearity in the concentration range of 0.1-1000 $ng.mL^{-1}$, and also indicate the same direction coefficient in both solutions. However, the results for Tl and Bi indicate that they are not suitable for use as internal standards for Br and I (Figure 6). Indium and Cs show similar behavior to that of Br and I and are considered suitable for use as internal standards.

Scattering of Data and Reproducibility

By the introduction of TMAH for washing, memory effects were reduced. Figure 7 shows the reproducibility of some elements determined in a 1% TMAH sample solution of mussel. For these elements, analysis was conducted 10 times (5 runs per sample) immediately after preparation of the sample solution. Between each analysis, the proposed new washing procedure was conducted to see if the memories of the target elements have been reduced to the background levels. The data show fairly good precision considering the low counts rates (Figure 7).

Detection Limits

The detection limits obtained are given in Figure 11,where two kinds of tendencies are observed ; (Group A and B), based on the ratio of TMAH counts/HNO_3 counts. A group, consists of the elements which have the ratio of over 1 and are clustered on the right side, and B group having the ratio of below 1, but more than 0.8 at lowest. Figures for ionization potential (given in parentheses) were calculated from the Saha equation. The data suggest that sensitivity is improved using TMAH relative to HNO .

In addition, Hg, I and Br may be lost during acid digestion and be clearly retained during TMAH digestion.

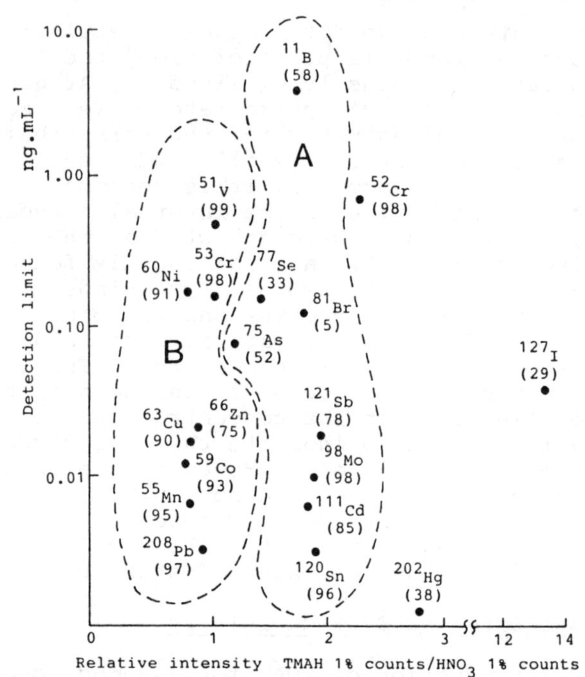

Figures in parenthses show calculated value for degree of ionization effciency

Figure 11 Relative intensity of TMAH 1% sol/HNO_3 1% sol versus detection limits

Determination of Br and I

Table 7 shows the analytical results of Br and I obtained from SRM sargasso, mussel, bovine liver and hair. Reference values for these elements are only available for the latter two SRMs. Therefore, Br and I contained in all of these samples were analyzed by radioactivation analysis to check the reliability of the TMAH-alkaline digestion method and its applicability to ICP-MS. These results are given in Table 7.

As observed, the values obtained by ICP-MS were higher by 5-10% than those of radioactivation analysis. However, since Br and I are not present in TMAH , it is unlikely that these high values are due to contamination.

So, the standard addition tests were made as shown

Table 7 Determination of Br and I in Standared Reference Materials with TMAH Digestion Method

n=5

Standard Reference Material	Internal Standard Elements	Br $\mu g.g^{-1}$		I $\mu g.g^{-1}$	
		ICP-MS	INAA	ICP-MS	INAA
Sargasso NIES NO.91	^{133}Cs	249.2±15.4	270±10	533 ± 6	520±20
	^{115}In	250.1±27.6		619 ±25	
Mussel NIES NO.6	^{133}Cs	90.1± 7.3	96± 1	4.96±0.11	4.2±1.3
	^{115}In	87.2± 2.6		6.00±0.78	
Bovine Liver NIST 1577a	^{133}Cs	10.3± 1.6	9*	0.38±0.03	0.31±0.30
	^{115}In	13.5± 0.8		0.38±0.04	
Hair NIES NO.5	^{133}Cs	72.1± 6.9	90*	0.83±0.06	0.62±0.08
	^{115}In	95.4±11.3		0.80±0.05	

* Reference Values only given

in Tables 8 and 9. The determined values quite well agreed with the added amounts of 100 $ng.mL^{-1}$ standard. Consequently, such large differences in the values determined for Br and I can not be reasonably explained to be the systematic error of this method.

Table 8 Analytical Results of Br and I in Sargasso (NIES NO.9) by Standard Addition Method (n=5)

Internal Standard Elements	Br		I	
	$\mu g.g^{-1}$	INAA	$\mu g.g^{-1}$	INAA
^{133}Cs	258±19	270±10	553±10	520±20
^{115}In	259±10		582±10	

In this study, it was observed that TMAH-alkaline digestion is effective for the determination of volatile Br and I since it does not cause any volatilization of the elements. It is interesting that the determination of such elements as Br and I, which has been nearly impossible by conventional methods, was now made possible

Table 9 Results of Br and I Obtained by Adding 100 ng.mL^{-1} Each to The Feed Solution of Sargasso

ng.mL^{-1}

Element	Internal standard elements	Concentration* in the feed solution n=5	Standard added to feed solution	Determined value n=5
Br	^{115}In	270±10	100	380±10
	^{133}Cs	280±20	100	380±10
I	^{115}In	610±10	100	720±20
	^{133}Cs	580±10	100	690±10

* Sample weight 1 mg / 1 mL 1% TMAH solution

by this method. This is particularly important for the determination of I in environmental samples.

SUMMARY

1. By using TMAH solution, it is possible to prepare a feed solution to the nebulizer without causing any losses of the elements which tend to volatilize by acidic digestion.

2. TMAH-alkaline digestion method can provide easy, safe and simultaneous treatment of many samples.

3. TMAH solution is quite effective for eliminating memories of such elements as Br, I, Hg, Na, Mg and others, which are difficult to remove by HNO_3 solution, and can give an excellent washing effect to lower such memories to background levels by use in a sequential washing procedure of ultrapure water - HNO_3 solution - TMAH solution.

4. It was found that, when determination of Br and I is made by using TMAH-alkaline digested sample solution on ICP-MS, ^{115}In and ^{133}Cs are suitable as the internal standard elements.

5. From the detection limits obtained, it was found that 1% TMAH solution can lower the detection limits of some elements such as I, Hg, Br, As, Cd and Sb, compared with 1% HNO_3 solution.

6. TMAH solutions do not generate any additional matrix effects different from HNO_3 and sensitivity is high. Indeed detection limits for Br and I are lower in TMAH solution.

7. When this method was applied to the determination of ^{81}Br and ^{127}I contained in several kinds of biological samples, the obtained values were slightly higher than those of radioactivation analysis and reference values.

ACKNOWLEDGEMENT

Sincere and hearty thanks to Dr. Tsurahide Cho, President of this company, who opened up the way to this study by inventing ultrapure TMAH, is acknowleged by the authors, thanks also for his valuable advice on many occasions in the course of this study.

REFERENCES

1. M.Mishima, 'Measurement of Trace Elements in Environmental Samples', Tokyo Kagaku Dojin, Japan, 1985, Chapter 1, p.20.
2. C.Veillon, Anal. Chem. 1986, 58, 1081. C.
3. J.Versiech and R. Corneilis, Anal. Chem. Acta., 1980, 116.
4. G. Knapp, Trace Element Analytical Chemistry in Medicine and Biology, Walter de Gruyter & Co., Berlin, New York, 1988, vol. 5, p.63.
5. S.Shimizu, T. Cho and Y.Murakami, Trace Element Analytical Chemistry in Medicine and Biology, Walter de Gruyter & Co., Berlin, New York, 1988, vol. 5, p.72.
6. S.Shimizu, I. Akabane, T. Cho and Y. Murakami, 'Nuclear Method in the Life Science' NIST. Gaitherburg, 1989, in printing.
7. W. K. Musker, J. Am. Chem. Soc., 1964, 86, 960.
8. A. L. Gray and A. R. Date, The Analyst, 1983, 108, 1033.
9. R. S. Houk, Anal. Chem. 1986, 58, 97A.

Re-Os ISOTOPE RATIO DETERMINATIONS BY ICP-MS: A REVIEW OF ANALYTICAL TECHNIQUES AND GEOLOGICAL APPLICATIONS

Jean M. Richardson[1], Alan P. Dickin[2] and Robert H. McNutt[2]

[1] Spectroscopy Geoscience Laboratories
Ontario Geological Survey
77 Grenville St., Toronto, Ontario,
Canada M7A 1W4

[2] Department of Geology, McMaster University,
Hamilton, Ontario, Canada L8S 4M1

1 INTRODUCTION

Rhenium (Z=75) is a dispersed chalcophile element, with two isotopes, which is chemically similar to Mo and W (Figure 1). In contrast, Os (Z=76) is a siderophile platinum group element (PGE) which has seven isotopes. Due to differences in their physical properties Re and Os are fractionated during natural processes, thus yielding a range in Re/Os ratios in geological materials. Both Re and Os are among the least abundant elements in the earth's crust and are typically found only at the $ng.g^{-1}$ level even in PGE and Mo ore deposits. In addition, ^{187}Re decays to $^{187}Os + \beta^- + \nu + Q$ ($T_{1/2}$ = 4.23 x 10^{10} Ma[1]) and this geochronometer has been used to determine the age and genesis of ore deposits, especially those containing Ni, PGE, Cu and Mo [e.g.2,3]. Studies have also included those of mantle evolution and crustal generation [e.g.4,5], origin of meteorites and cosmogenesis [e.g.1,6,7] or aspects of the geology of the Cretaceous-Tertiary boundary[8].

In the past, the high ionization potential of Os (IP = 8.5eV) impeded the widespread use of the Re-Os geochronometer. Indeed, Re (IP = 7.87 eV) is so poorly ionized that it is used to make the filaments from which other elements are vaporized during thermal ionization mass spectrometry. Recently, Re and Os have been determined by other mass spectrometric techniques including resonance ionization mass spectrometry (RIMS)[9,10], secondary ion mass spectrometry (SIMS)[11-13], accelerator mass spectrometry (AMS)[14] and laser microprobe mass analysis (LAMMA)[15].

Re-Os abundances and their isotopic ratios in geological materials have been successfully determined by inductively coupled plasma-mass spectrometry (ICP-MS). This paper reviews the pertinent chemical and instrumental techniques, assesses the precision and

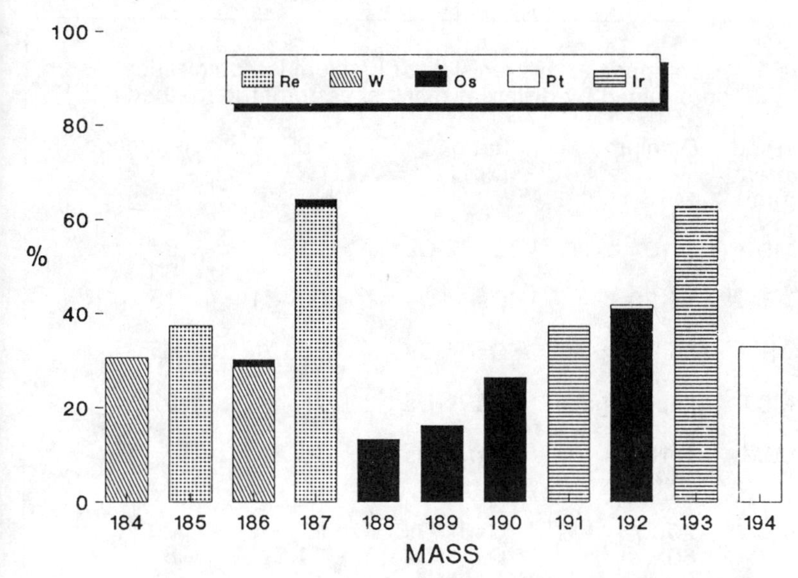

Figure 1 Isotopic abundances in the range 184 to 194 amu.

accuracy and compares these figures of merit to other mass spectrometric techniques. The contribution of this data to the understanding of geological processes is also reviewed. Finally, possible future developments in chemistry and instrumentation that could improve Re and Os determinations are outlined.

2 DEVELOPMENT OF ANALYTICAL TECHNIQUES FOR Re AND Os DETERMINATION BY ICP-MS

The combination of an inductively coupled argon plasma ion source and a quadrupole mass spectrometer resulted in the development of inductively coupled plasma - mass spectrometry (ICP-MS)[16,17]. The temperature of an argon plasma is sufficiently high to ionize 78% of the Os and 93% of the Re in a sample[18] and the quadrupole mass spectrometer provides the mass resolution necessary to separate the various isotopes. Manufacturers of ICP-MS instruments have quoted Re and Os detection limits in the ng.mL^{-1} range. A comparison of published precision data indicate that ICP-MS is competitive with the other MS techniques for the measurement of Re and Os isotopic ratios (Table 1) and the technique is therefore a viable alternative to RIMS, SIMS and AMS for Re and Os determinations.

Table 1 A comparison of precision of osmium isotopic ratios determined by different mass spectrometric methods

Method	Osmium content (ng.mL^{-1})	Ratio (2σ)	RSD %	Reference
SIMS	5-10	$^{187}Os/^{186}Os$	1 - 3	64
				63
	15-20	$^{187}Os/^{186}Os$	0.5 - 1	13
AMS	55	$^{187}Os/^{186}Os$	20	65
RIMS	ppb range	$^{187}Os/^{186}Os$	1 - 5	9
LAMMA	1-100	$^{187}Os/^{186}Os$	1 - 2	7
ICP-MS				
SN	2000	$^{187}Os/^{186}Os$	2	20
	60	$^{187}Os/^{186}Os$	14.2	8
	(a)	$^{187}Os/^{188}Os$	1.6	21
	400	$^{187}Os/^{186}Os$	1	24
OG	< 5	$^{187}Os/^{192}Os$	2	15
	0.1	$^{187}Os/^{192}Os$	1 - 2	1
		$^{187}Os/^{190}Os$		1
	10	$^{187}Os/^{188}Os$	1	22a
	(b)	$^{187}Os/^{188}Os$	9.5	23
	100 - 400	$^{187}Os/^{188}Os$	0.34	30
	(a)	$^{187}Os/^{188}Os$	3	21
ETV	(a)	$^{187}Os/^{188}Os$	3.75	21
MM	0.8	$^{187}Os/^{186}Os$	10	24

(a) average of 4 determinations on 50 mg iridosmine (Ir-Os alloy).
(b) isotopic spike
SN solution nebulization
OG Os generator
ETV electrothermal vaporization
MM microheater/merging

The first Re and Os isotopic measurements made by ICP-MS were reported by Lindner *et al.*[7] for Re and Lichte *et al.*[8] for Os. Date *et al.*[19] reported the first ICP-MS determinations of platinum-group elements (PGE) (excluding Re and Os) in the geological reference materials SARM 7 (South Africa Bureau of Standards, Pretoria, R.S.A.) and PTC 1 (CCRMP, CANMET, Ottawa, Canada.). Although Os can be determined using pneumatic nebulization[8, 20-21], development[22] and modification of an "Os generator"[15, 22a-23] has allowed the on-line distillation of Os from a sample and thus determination in the gaseous form. This technique significantly improves sample yield and analytical precision. Sample introduction techniques such as microheater/merging[24] and electrothermal vaporization (ETV)[21] have also been investigated and proved useful for low level determinations.

Before geological problems could be investigated using Re-Os isotopes by ICP-MS, two aspects of the overall analytical technique needed extensive developmental work. First, those aspects specifically associated with ICP-MS had to be evaluated, *i.e.* the occurrence of matrix effects, isobaric overlap, signal optimization and memory effects. Second, the broader considerations associated with the chemical preparation of geological samples containing Re and Os had to be addressed. Many of the techniques used to prepare samples for PGE determinations[25,26] can be adapted, although the extreme volatility of Os has precluded its determination by these techniques in some cases[19,27]. In addition, the differences in chemical affinity between Re and Os are such that some sample preparation techniques which are successfully used for PGE determinations are not compatible with Re sample preparation.

3 SAMPLE PREPARATION FOR ICP-MS DETERMINATIONS OF Re AND Os

Potential limitations

The preparation of rock or mineral samples for Re and Os determinations must take into account several potential problems (a) the ready oxidation of Os to OsO_4, (b) the low abundances of Re and Os in natural materials, (c) the sporadic occurrence of small trace minerals with high Re or Os contents in larger masses of material which are impoverished in Re or Os (nugget effect), (d) the relatively insoluble nature of Os-PGE alloys and the presence of Re and Os in both silicate and sulphide minerals in a given sample. Furthermore, most geological material must be dissolved prior to analysis by ICP-MS, yet strongly acidic solutions high in dissolved solids are difficult to analyze by this technique. Many of these geological and chemical considerations also impact on the successful determination of PGE by other analytical techniques, suggesting that it should be possible to use previously developed sample preparation procedures (*e.g.* direct dissolution, fire assay, alkali fusion[25-26]) with some modifications.

Fire assay

Not all rocks and minerals will dissolve in inorganic acids. Furthermore, although fusion techniques will dissolve Os-PGE alloys, these are extremely oxidizing, high temperature (*i.e.* $<130\,^{\circ}C$) techniques and only relatively small samples can be accommodated.

The fire-assay technique of Robert *et al.*[27] concentrates most metals in a 50 g powdered sample into a solid, 20 g NiS button, thus minimizing the nugget effect. Osmium is effectively concentrated in the button[27,28], but Re is partitioned between the flux (90%) and the button (10%)[13]. In addition, this technique homogenizes all metals present in the silicate and sulphide phases of the sample into a single button. The technique has been modified to use less Ni but the sample is fused for a longer time interval (16 h) to yield a smaller (1-3 g) button (R. Keays, pers. comm. 1989). This procedure shortens the crushing time, decreases the volume of reagents required and lowers blank levels. The longer fusion time is necessary to ensure that metal recovery from the silicate, sulphide and/or alloy minerals present in the sample is complete.

Richardson *et al.*[29,30] discussed the advantages and disadvantages of the fire assay techniques with specific reference to Re isotope determinations by ICP-MS. They concluded that although such techniques can be used, matrix elements (Fe, Cu, Ni, Na *etc.*) in the resultant solutions degrade the quality of the analytical data and must be removed by ion-exchange chromatography or precipitation to obtain high precision Re isotope determinations.

Solid spike

The ready oxidation of Os to gaseous OsO_4 above 130°C has been a major barrier to routine quantitative Os determinations. Coupled with the low abundances in most geological materials, this has meant that Os is not typically determined with the other PGE[19,27]. A solid-solution isotopic spike[23] was prepared to facilitate quantitative Re and Os abundance determinations by isotope dilution.

Such a solid spike has several advantages over traditional liquid spikes. First, being in a solid form, it can be stored with no loss of Os. This is often not the case with liquid solutions containing Os [e.g.6]. Second, the matrix is NiS and thus is in the same chemical form as many minerals (*e.g.* pyrrhotite $[Fe, Ni]_{1-x}S$). This means that if the spike is crushed to the same grain size as the sample, the behaviour of both should be similar during dissolution or fusion. A solid spike is particularly useful during the fire-assay technique, because artificially enriched Os in a liquid spike could be preferentially volatilized before homogenizing with natural Os in the sample.

Sample preparation procedure

A schematic flow chart illustrating sample preparation techniques for Re-Os isotopic determinations of rocks or minerals by ICP-MS is shown in Figure 2. Details of the procedure are given elsewhere[31]. Typically, samples are weighed, spiked, dissolved or fused, the sample and spike equilibrated and the non-analytes removed. To minimize sample handling and to enhance the signals

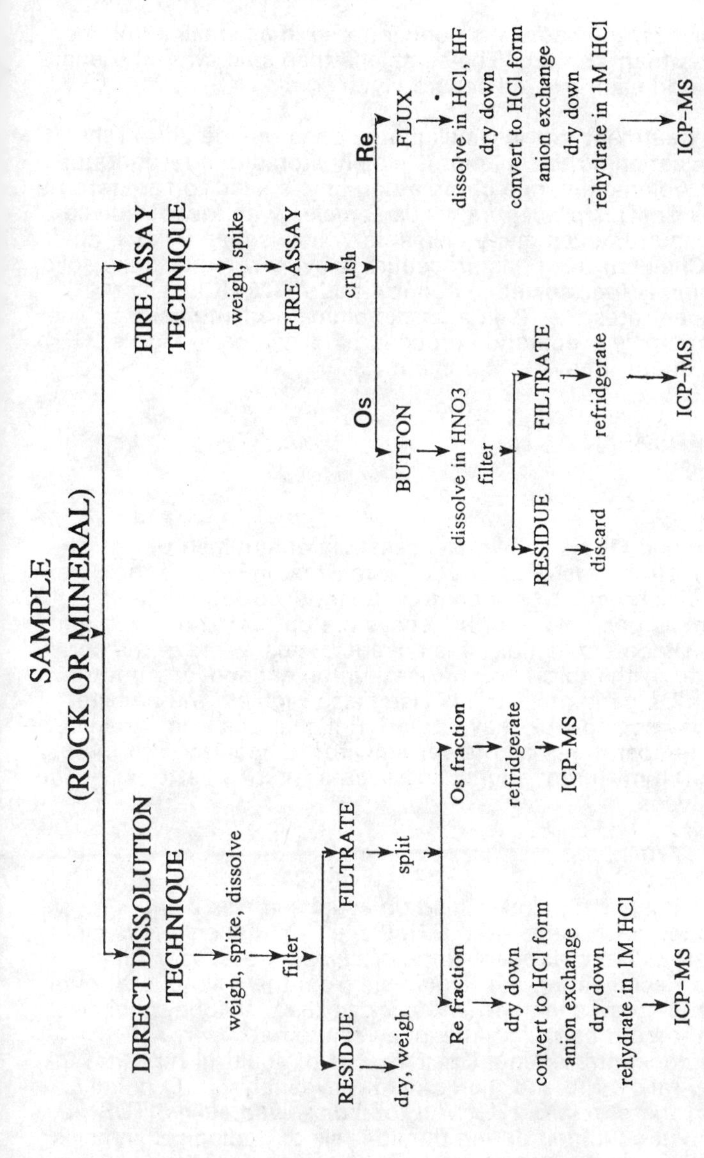

Figure 2 Schematic sample preparation flow chart

during the analyses, the sample is concentrated in as small a volume as possible (less than 25 mL). The sample is then split into a Re and an Os fraction and each treated separately.

Further treatment of the Os aliquot is done on-line during the ICP-MS determination so this aliquot is simply stored in a refrigerator until analyzed. Anion-exchange chromatography is used to separate Re from the matrix[30]. This procedure yields samples with low dissolved solid contents and removes many elements which isobaricly overlap Re. The 8M HCl eluent from this procedure is evaporated and the solid residue remaining is redissolved in about 4 mL of 1M HCl. This procedure concentrates the Re in a small volume of dilute acid[30]. The total, the total sample preparation procedure delineated in Figure 2 typically takes about 5 days to complete.

4 EXPERIMENTAL

Instrumentation

Rhenium and Os have been successfully determined using ICP-MS prototype instruments[7,15], the VG PlasmaQuad[1,20,24] and the Sciex Elan[8,21-23]. Typically, mass-flow controllers are used to precisely control the sample gas-flow and the lenses are optimized for high mass elements. For Os determination, it is advantageous to move the spray chamber outside of the torch box facilitating the changeover from the nebulizer to the Os generator[23]. This also helps regulate the ambient temperature. Davies and Du[32] have reported higher electron densities, higher plasma temperatures and fewer polyatomic species with the use of a "long" (150 mm) torch, suggesting such a torch is better suited to Re and Os analyses.

Determination of rhenium

Rhenium is typically determined on liquid samples using conventional pneumatic nebulization. However, this technique is highly inefficient and thus controls the volume of sample necessary for a determination. Recirculating the Re sample from the test-tube through the nebulizer and spray chamber, and back to the test-tube, minimizes the impact of the poor sampling efficiency of the nebulizer. A Re sample can be concentrated in a small volume of liquid (4 mL), run for 15-20 minutes, recovered and then stored for reanalysis. Secondly, the presence of more than 0.1% (w/v) total dissolved solids (TDS), a relatively common condition during the analysis of geological samples can attenuate the count-rate and enhances instrument drift (*i.e.* Figure 3) and degrades precision. The diminished count-rates are attributed to sample deposition on the sampler and skimmer [e.g.33]. Douglas and Kerr[34] indicate that a wash-period four times longer than the determination time is necessary for a solution containing 1.0% TDS. After the anion-exchange technique, the washout time after a Re determination is negligible (20-30 seconds). In addition, the presence

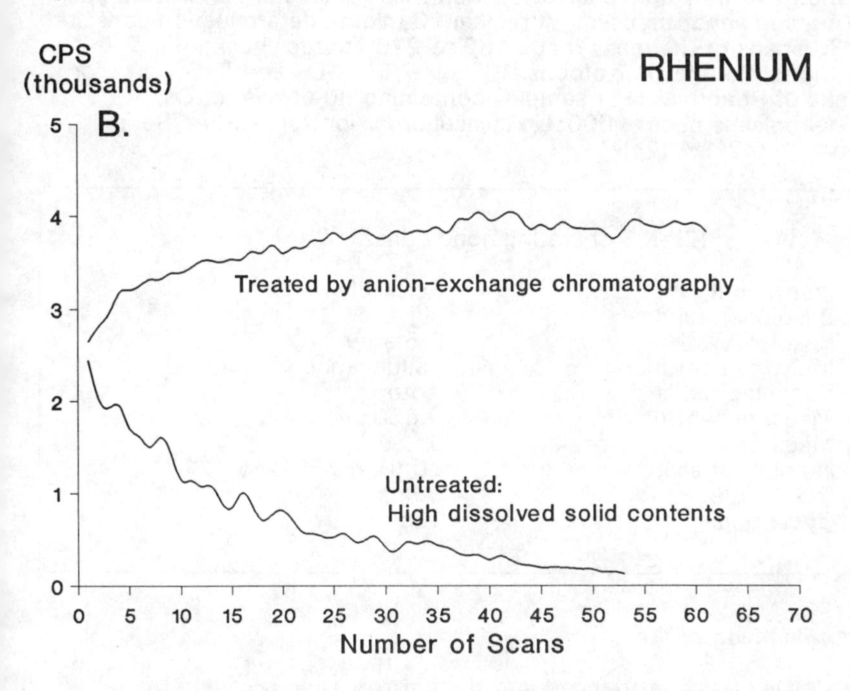

Figure 3 A run profile (count rate vs. time) for Re. The rhenium count-rate is greater and more stable in samples that are treated by anion-exchange chromatography to remove the high dissolved solid contents.

of a single matrix element at high concentrations can lead to analyte signal suppression. The magnitude of the suppression has been shown to increase with the mass of the matrix element [e.g.35-36]. The Re profile shown in Figure 3 indicates that a higher, invariant count-rate can be generated from a sample from which the matrix elements at high concentration levels (Fe, Ni or Cu derived from the sample or Na and B added during sample preparation) are removed by anion-exchange chromatography. This technique has the advantage of both removing the matrix and concentrating the analytes in dilute acid.

Optimum ICP-MS operating conditions are shown in Tables 2a and 2b. To make the data-collection parameter set as flexible as possible (*i.e.* use the same set of conditions for mineral, rock and fire-assay analyses), several background positions should be collected (*e.g.* 181, 183 and 203). Richardson *et al.*[23] monitored 217 and 219 amu, masses that are not characterized by naturally-occurring elements or polyatomic species. Half-mass positions suffer from peak tailing so cannot be used. A typical 20-minute determination consists of 80 scans of the mass range 180 to 220 amu for background positions, Re and Os isotopes (^{185}Re, ^{187}Re, ^{190}Os and ^{192}Os) and for peaks of Pt and W. For samples containing 30 ppb Re or Os, signal/noise is about 1000:1. Typical precision for $^{185}Re/^{187}Re$ is between 0.3-1% (2σ)[37].

Table 2a ICP-MS operating conditions

Instrument	Sciex Elan
Resolution mode	high
Resolution	0.6 amu
Measurement mode	multichannel
Scanning mode	isotopic
Measurement time	0.060 sec
Measurements per peak	5
Number of scans	80 (Re), 80(Os)
Dwell time	60ms
Cycle time	10s

Determination of Os

Os can be introduced into the plasma using a variety of techniques. Gregoire[21] has used pneumatic nebulization and an electrothermal vaporization device with a graphite platform (ETV). More commonly, the easy oxidation of Os to gaseous OsO_4 is utilized to distill this analyte away from interfering isobaric species using on-line devices such as an "OsO_4 generator"[22] or a "microheater/merging" apparatus[24]. The latter, consists of a small glass cup in a glass ampoule (the merging chamber) which is mounted on a microheater. This assembly is inserted upstream of the torch in the sample gas-flow line. OsO_4 is released from the sample into the sample gas as heat is applied. In both techniques, an Os-bearing sample is combined with a strong oxidizing agent in a glass chamber through which the sample gas-flow is directed (Figure 4); gaseous OsO_4 is entrained in the sample gas-flow, delivered to the plasma and ionized.

Os generator The OsO_4 generator of Bazan[22] consists of a glass sampling chamber fitted with a cooled glass reflux condenser placed on-line in the sampling gas flow. The sample is injected through a port and retained above a porous gas frit by the pressure of the sampling gas flow. A strong oxidizing agent is added to the sampling chamber and OsO_4 is immediately released and entrained in the sampling gas flow (Figure 4). The Os generator has been used to produce Os

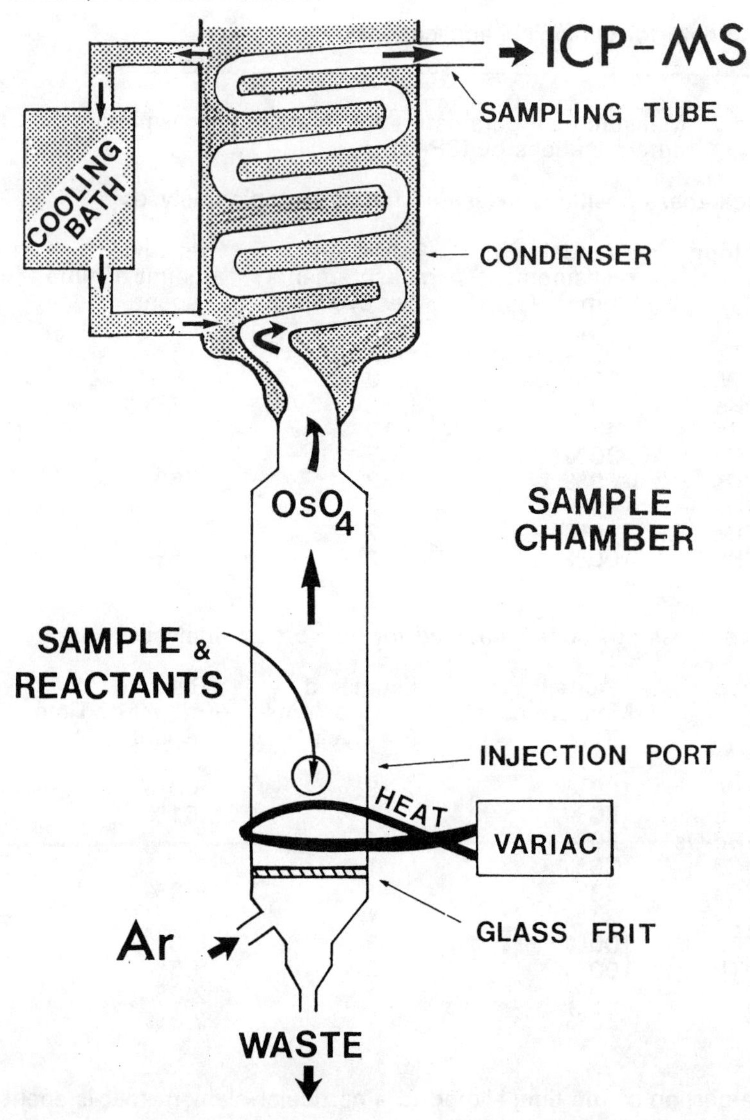

Figure 4 Diagram of the osmium generator[23]

isotopic data for geological applications[1,7,38].

Table 2b Measurement parameters for rhenium and osmium determinations by ICP-MS.

Typical mass positions measured for Os isotopic analyses.

Isotope	Actual measurement time*	Estimated measurements per cycle	Relative amount of time spent
^{181}Ta	100%	2	6%
^{183}W	100%	2	6%
^{185}Re	100%	2	7%
^{187}Re-Os	43%	10	
^{188}Os	30%	7	
^{189}Os	9%	2	75%
^{190}Os	9%	2	
^{192}Os	9%	2	
^{194}Pt	100%	2	6%

Typical mass positions measured for Re isotopic analyses.

Isotope	Actual Measurement Time*	Estimated Measurements Per Cycle	Relative Amount of Time Spent
^{183}W	100%	3	8.3%
^{185}Re	50%	11	61%
^{187}Re-Os	50%	11	
^{189}Os	33%	1	
^{190}Os	33%	1	8.3%
^{192}Os	34%	1	
^{194}Pt	100%	3	8.3%
^{203}Tl	100%	3	8.3%
217	50%	1	2.8%
219	50%	1	2.8%

*The portion of the time alloted to a particular element that is spent measuring the given isotope.

The use of the Os generator has a number of advantages over other methods of sample introduction. The temperature-sensitive oxidation reaction of Os to OsO_4 can be easily manipulated. Thus, for a given amount of sample, generator-mode determination yields a count-rate two orders of magnitude higher or alternatively, uses 100 times less sample to achieve the same signal as conventional solution nebulisation. In addition, a small portion of the sample can be run to exhaustion and all the Os delivered to the plasma Another significant advantage of on-line distillation technique is that the OsO_4 delivered to the plasma is purged of most matrix and isobaric interferents. Re forms an oxychloride and several bromides and fluorides that are slightly volatile below 200°C. To avoid isobaric interference from ^{187}Re on ^{187}Os during Os analyses, HF and HBr are excluded from the

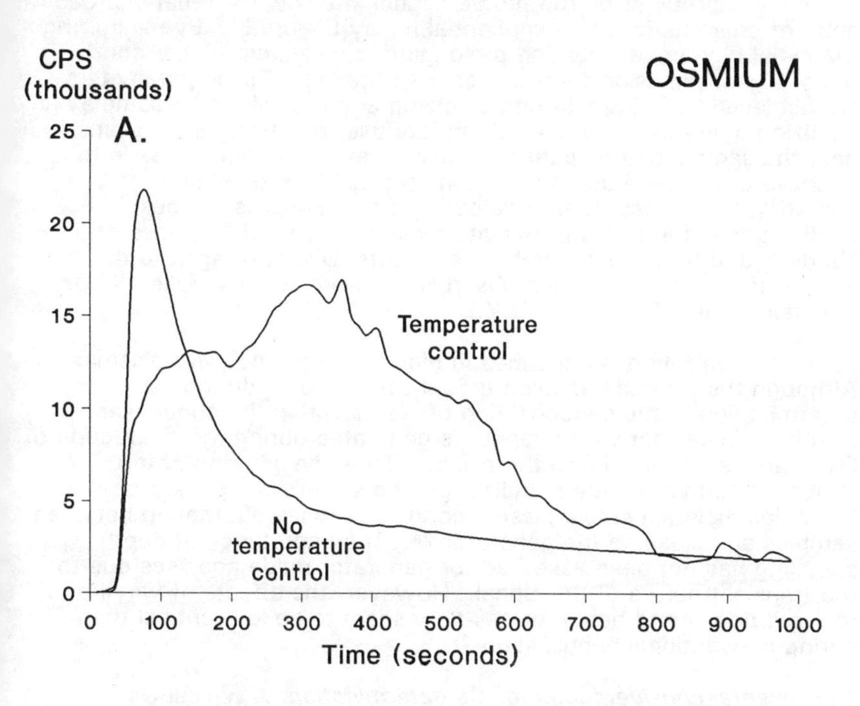

Figure 5 The run profiles (count rate vs. time) for Os. The osmium signal is transient, increases exponentially and decays quickly if the temperature of the reaction is not regulated. A similar signal response with time is generated for samples introduced into the instrument by Os generator, ETV or microheater/merging techniques. Temperature control yields a slightly lower count-rate which persists longer.

sample preparation techniques. HCl is also avoided if possible. Otherwise, low temperature volatile transition metal species do not form so Pt, Ni, Cu, Zr, W and the remainder of the matrix remain in the sample chamber and are later expelled as waste. Thus chemical pretreatment of the sample is limited and the potential for Os loss is reduced. Finally, the sample can be recovered and analysed for other analytes if desired.

In contrast to solution nebulization, the samples produced by Os distillation techniques result in a transient signal of variable length that drops rapidly (Figure 5). The rate of OsO_4 generation can be controlled by the choice of oxidizing agent and by regulating the temperature at which the determination is conducted. Counting statistics improve if the run profile (signal intensity vs. time) is broader and not characterized by exponential decay (Figure 5). By minimizing the initial high count-rate and prolonging subsequent higher count-rates, higher precision Os analyses are obtained. These parameters are achieved by using a strong oxidizing agent, chilling the sample and oxidizing agent until immediately before use, and using a heat tape to heat the sample and reagent immediately after the initial peak in the count rate has passed. Optimization studies by Bazan[22] and Richardson[23] suggest that 10% periodic acid H_5IO_6 is the best oxidizing agent in terms of oxidation capability, shelf life, ease of handling and lack of concomitant elements. Disadvantages to this acid include the presence of Re or Os in H_5IO_6 from some suppliers (*e.g.* Eastman Kodak, Rochester, N.Y.).

Os ionization is enhanced in higher energy, hot, dry plasmas. Although the plasma has been thought to be "dry" during Os determination[15], the condensation of water within the condenser spirals indicate that water vapour is generated during the production of OsO_4 and is entrained into the plasma. Thus the addition of a "bubbler" to hydrate the nebulizer gas-flow while the generator is off-line helps maintain stable plasma conditions when alternating between samples and cleaning the generator[22a]. The importance of depth sampling has not been assessed for generator-mode analyses due to the transient nature of the signal. However, the effects of varying forward power and nebulizer gas-flow seem to be less critical than during conventional nebulization[22a].

Instrumental considerations for Os determination A typical Os determination consists of analyzing about five 1 mL aliquots of a sample using a parameter set similar to that in Table 2b. Five isotopes of Os are collected (192, 190, 189, 188, 187) and two low and one high mass positions to monitor background (^{181}Ta, ^{183}W and ^{194}Pt). ^{185}Re is collected to monitor for Re interference on the ^{187}Os peak. Forty to eighty scans are collected in 10 - 15 minutes, after which the Os complement of the aliquot is generally exhausted. A complete determination takes 60 - 90 minutes. Average precisions of $\pm$ 0.34% RSD on $^{187}Os/^{188}Os$ are possible on geological materials.

The most significant isobaric overlap that interferes with Re-Os determinations occurs at 187 amu. This is the radioactive isotope of natural Re (62.6%) and the minor radiogenic isotope of Os (1.64%). The low abundance of ^{187}Os means that counting statistics for this isotope limit the precision of Re-Os isotopic determinations[22a] . Different procedures can be used to minimize this interference in either Re or Os analyses.

If the anion exchange procedure is used prior to Re determinations, Os and Pt are retained on the columns this minimises the ^{187}Os interference on ^{187}Re[30,37]. The small resulting intererence, if any, can easily be corrected for using ^{190}Os/^{188}Os determined during Os generator-mode determination. In contrast, the on-line distillation of OsO_4 in Os determinations results in most of the Re present in an Os aliquot being retained in the generator and expelled as waste. However, ^{185}Re is monitored during Os determinations to allow a ^{187}Re correction on ^{187}Os. Typically, such a correction is small. Because the Os isotope ratios are determined on a sample spiked with a mixed ^{185}Re-^{190}Os spike (to allow isotope dilution determinations), ^{185}Re/^{187}Re is enhanced in all determinations, thus improving the precision of the ^{187}Re correction on ^{187}Os. If necessary, the correction process can be iterative for both Re and Os.

There are also direct overlaps at 184 amu between Os, W and $^{92}Zr_2$, at 186 amu. between Os, W and $^{93}Nb_2$, at 188 amu. between $^{94}Zr_2$ and at 190 and 192 amu between Os and Pt. The probability of such interferences is enhanced by geochemical similarities between Os and Pt, Re and W, and the use of Zr crucibles in sample fusions[20]. Fortunately, the W interferences on Os at 184 and 186 amu are irrelevant because these masses are not required for Re or Os isotopic analyses. The on-line distillation of Os from the sample using the generator results in interferents such as Pt, Zr and W remaining in the sampling chamber, thus negating their importance.

Other polyatomic species and oxides present are limited due to the high, narrow mass range under consideration (185 - 194 amu), the separation of the analytes from the matrix using either anion exchange or distillation and the physical properties of Re and Os. The narrow, high mass range under consideration means that the formation of sulphides, chlorides or argides derived from the matrix is minimal. Yb, Lu, Tm and Hf oxide species can potentially interfere. However, because these compounds are typically of lower abundance and have lower diatomic bond strengths than geochemically similar Dy and Zr oxides, the formation of the interfering oxides can be monitored using DyO and ZrO. The very high second ionization energies of Re (16.6 eV) and Os (17.0 eV) make the formation of doubly charged ions of these elements unlikely.

Memory effects These effects have been reported during Re-Os analyses[15,20-22]. These are caused by several factors. First, Os tends to be retained on glassware and diffuse into plasticware (especially Teflon). Second, samples that have high dissolved contents can be viscous and are thus easily retained. Third, the high vapour pressure of OsO_4 means that gaseous Os could deposit solid Os compounds on lower temperature surfaces (*e.g.* samplers, skimmers, photon stops) or fill the sample introduction devices or lines and dissolve into subsequent samples[21]. Preventative maintenance to avoid memory effects and sample cross-contamination includes short-term use of disposable plasticware, avoidance of viscous samples, rigorous cleaning of glassware (especially torches) after use and the insertion of blanks between samples.

Os determinations using the OsO_4 generator can suffer from two types of memory. Sample can be either retained in the generator or plated on the torch, sampler, skimmer or ion optics. Unfortunately, due to the nature of the generator construction and support gas-flow, it is difficult to clean the apparatus while on-line. Bazan[22] designed a top-down "flushing arrangement" for the generator, but there was no facility to replace the generator without shutting down the instrument. The "bubbler bypass" shown in Dickin *et al.*[22a] (Figure 1 in ref 22a) allows the generator to be taken off-line without shutting down the instrument. In this case, the generator can be either replaced or filled to the top stopcock with a strong oxidizing agent and allowed to stand for a short period of time. Rinsing with distilled water[21] is not sufficient.

After either replacement or "on-line" cleaning of the generator, another sample can be analyzed if the count-rate at any mass between 185 and 192 is less than three times background. If not, the instrument must be shut down, the torch changed and the sampler cleaned. Excessive osmium plating on the skimmer or ion stops has not been reported and can apparently be avoided by routine maintenance. Torches used for Re and Os determinations must be changed daily and, like the tetroxide generators, cleaned in aqua regia or an emulsifying agent like RBS35 (Pierce Chemicals, Rockfort, Ill.). The easy oxidation of Os is an obvious advantage during cleaning.

5 RESULTS

Data handling

A typical Os determination generates over 11,250 measurements and averages about 340 kb in computer file-size. Data-handling routines for Re and Os analyses are complex, but are documented in Dickin *et al.*[22a] and Lindner *et al.*[1]. Data reduction consists of calculating the maximum peak height at a given mass, correcting for isobaric interference and mass bias using ^{188}Os. Dead-time corrections must be used if count-rates exceed a million cps. Further statistical treatment utilizes unpublished data handling routines modified from those used in TIMS (McMaster University) or the ISACS software of Hudson (pers. comm. in[1], Lawrence Livermore National Laboratory, University of California).

Traditionally, Os isotopic results are reported with respect to 186 amu (*i.e.* $^{187}Os/^{186}Os$). However, as pointed out by Masuda *et al.*[20], ^{188}Os is a better normalization choice for several reasons. First, ^{188}Os (13.27%) is more abundant than ^{186}Os (1.59%) and thus easier to measure precisely. Second, ^{186}Os has an isobaric overlap with an abundant W isotope (^{186}W, 28.64%). Third, ^{190}Pt (0.0127%) decays to ^{186}Os with a half-life that is geologically meaningful ($T_{½}$ 5.9 $\times 10^{11}$y). In most geological materials, the contribution this decay makes to ^{186}Os is negligible, but in Pt-rich minerals, this addition could be significant. To convert $^{187}Os/^{186}Os$ to $^{187}Os/^{188}Os$, the former ratio must be multiplied by the $^{186}Os/^{188}Os$ ratio (0.12035), which is constant in nature.

Reference materials and standards

Isotopic measurements are typically compared to standards (*e.g.* NBS 978 for Sr, NBS 971 for Pb). Unfortunately, there is currently no inventory of Os isotopic compositions which can be used in a similar fashion. However, the ratios of the stable isotopes of Os are constant in nature and the isotopic composition of Os was determined by Nier[39] and later by Luck and Allègre[6]. ICP-MS determinations by others[15,20,22a] agree well with the earlier values (see Table 3).

Table 3 Abundance of the isotopes of osmium. All are normalized to the most abundant isotope, ^{192}Os.

Os isotope ratioed to ^{192}Os	1	2 SIMS	3* RIMS	4	5
	%RSD	%RSD	%RSD	%RSD	%RSD
^{190}Os	64.4 (1)	64.4 (0.09)	64.04 (0.5)	64.21 (0.22)	64.43 (0.22)
^{189}Os	39.3 (1)	39.3 (0.06)	40.48 (0.5)	40.00 (0.10)	39.55 (0.26)
^{188}Os	32.4 (1)	32.44[b]	32.47 (0.5)	32.37 (0.37)	32.44[b]
^{187}Os	3.93 (3)	3.69 (0.11)	3.66 (0.5)	N.R.[a]	3.79 (0.8)
^{186}Os	3.88 (3)	3.90 (0.10)	3.87 (0.5)	3.88 (0.52)	3.84 (0.7)
^{184}Os	0.05 (10)	0.06 (20.34)	N.R.	0.04 (9.34)	0.05 (8.4)

a. Natural variation expected
b. $^{192}Os/^{188}Os$ used for normalization
N.R. not reported
* not corrected for fractionation
( ) 2 sigma uncertainty.

1. Reference[39]
2. Reference[6]
3. Reference[9]
4. Reference[20]
5. Reference[15]

Gramlich *et al.*[40] determined $^{185}Re/^{187}Re$ to be 0.59738 ± 0.00039. ICP-MS determinations by Richardson *et al.*[30] indicate this ratio can be determined to a precision of ± 0.5% (2σ RSD) following anion-exchange chromatography (Table 4).

The measurement of isotopic ratios in solutions of pure spike can be used to determine the accuracy of ICP-MS isotopic measurements because a Certificate of Analysis is supplied when artificially enriched isotopic materials are purchased. The Re-Os-Ni solid-solution spike prepared by Richardson *et al.*[23] has a $^{190}Os/^{192}Os$ ratio of 51.5 ± 0.9, a $^{187}Os/^{188}Os$ ratio of 0.063 ± 0.006 and

Table 4 Isotopic compositions and elemental abundances for spiked samples of pyrrhotite from the Creighton Mine, Sudbury, Canada[38]

Osmium

$^{187}Os/^{186}Os$ (2σ)	%RSD (2σ)	$ng.g^{-1}$
0.7810 ± 0.0043	0.41	112.7
1.0638 ± 0.0023	0.22	115.1
0.8320 ± 0.0040	0.47	125.7
0.7300 ± 0.0015	0.21	134.6
1.0790 ± 0.0068	0.63	143.7
1.0713 ± 0.0041	0.39	155.1
0.7380 ± 0.0043	0.58	341.4
0.7757 ± 0.0010	0.13	368.8

Rhenium

$^{185}Re/^{187}Re$ (2σ)	%RSD (2σ)	$ng.g^{-1}$
2.0233 ± 0.0032	0.16	155.2
1.4652 ± 0.0010	0.65	193.5
1.6448 ± 0.0067	0.41	199.5
1.9609 ± 0.0080	0.41	207.3
1.6058 ± 0.0066	0.41	310.5
1.4341 ± 0.0071	0.50	354.4
1.1909 ± 0.0031	0.26	456.2
1.2256 ± 0.0027	0.28	558.4

$^{185}Re/^{187}Re$ ratio of 17.04 ± 0.41 (2σ), values that are accurate within the precision quoted by the manufacturers[23].

6 APPLICATIONS OF Re AND Os DETERMINATIONS BY ICP-MS

Half-life of ^{187}Re

Although Naldrett and Libby[41] discovered that ^{187}Re was a long-lived beta emitter, attempts to precisely determine the half-life of this decay interval have been frustrated by the very low E(max) (2.6 keV) of the beta distribution. Early determinations are in "rather poor agreement"[1] and were based on either determination of Re and Os in minerals[42], meteorites dated by other techniques[6], or direct counting experiments[43-44]. Recent determinations by direct counting, double spike, isotope dilution ICP-MS analyses of five large samples of initially almost Os-free Re has resulted in a new precise determination of the half-life of Re ($T_{½}$ = 4.23 + 0.13 x 10^{10} y)[1]. This value supersedes the previously accepted value of Lindner *et al.*[7]

Cosmogenesis

The work by Lindner *et al.*[1] on the half-life of ^{187}Re has generated new insights into the formation of the Galaxy and Solar System. Following the approach of Clayton[45] ,Luck and Allègre[6] used a value for the Re decay constant derived from meteorites to calculate the age of the Galaxy to be between 10,600 and 16,800 Ma. The agreement at the 2σ level between this value and that determined by Lindner *et al.*[1,7] from ICP-MS determinations has meant that the age of the Galaxy cannot be more precisely calculated.

However, using the more precise half-life, a difference in the closure time for ^{187}Re in iron meteorites with respect to chondritic meteorites was ascertained. This difference could have arisen in two ways, both of which provide strikingly new observations about the formation of the Solar System. Either there was a time delay (380 ± 160 Ma) between the formation of the iron meteorites and that of chondritic meteorites, with the iron meteorites being younger, or there was variation in the initial ^{187}Os ratio in the early solar nebula[1] [using Lindner *et al.*'s decay constant]).

Ore genesis

Although Re-Os isotope determinations obtained by other techniques have been applied to ore genesis[13,46], the results of Richardson *et al.*[38] on the Sudbury Ni-Cu-PGE complex are the first Re-Os determinations by ICP-MS to evaluate ore genesis.

Although the genetic processes responsible for the formation of the 1850 Ma Sudbury Intrusive Complex[47] are equivocal (meteoritic impact[48-49], or endogenic event[50]), the geology and geochemistry of the intrusion is well documented. Field, major and trace element data and Rb-Sr and Sm-Nd isotopic determinations indicate that the igneous rocks of the Sudbury Intrusive Complex assimilated significant amounts (at least 50%) of pre-existing heterogeneous crustal material[51-54]. Precise but preliminary Re-Os isotope determinations on ore minerals illustrate the range and radiogenic nature of the initial Os ratio at Sudbury, compared to the mantle growth line and crustal evolution envelope of Allègre and Luck[4] (Figures 6 and 7). The radiogenic nature of these data indicates that the ore, like the silicate rocks, did not originate solely from juvenile mantle or meteoritic material but appear to have involved large amounts of crustal material from several sources.

Cretaceous-Tertiary Boundary

Alverez *et al.*[55] proposed that the mass extinctions at the end of the Cretaceous Period were caused by the impact of a large (10 km) meteorite which left anomalously high Ir levels in sediments that were being deposited at that time. The geochemical coherence of Ir and Os and the difference between the Os isotopic ratio of meteoritic material (well characterised by Luck and Allègre[6]) and crustal rocks at 66 Ma indicated that Re-Os isotopic determinations could help clarify the origin of Ir in these shales. Lichte *et al.*[8] confirmed that the Os (and by inference Ir) in the Woodside Creek, New Zealand shales is extra-terrestrial in origin.

7 FUTURE IMPROVEMENTS IN ICP-MS DETERMINATIONS OF Re AND Os

Advances or variations in sample preparation techniques could simplify the procedures outlined here. Chelex resin has been used for Os purification[13] and molecular recognition products such as Superlig™ (IBC Advanced Technologies, Orem, Utah) are available but not widely used. The possibility of analyzing ion-exchange resin directly by either nebulizing a slurry of resin or injecting the resin directly into the generator could modify sample preparation techniques. Alternatively, HPLC-type prepacked columns could combine sample preparation with on-line analysis. The production of OsO_4 off-line via a cold-finger arrangement and development of a glass-cracker assembly similar to that used in stable isotope determination might also be possible. Os can also be distilled into a receiving acid or solvent and nebulised[8,10,21]. Re also can be concentrated by solvent extraction. Finally, the possibility of direct measurement of isotopic ratios of minerals or fire-assay buttons by laser ablation ICP-MS has not yet been investigated.

Flow injection analysis (FIA[56]) is a sample introduction technique that shows promise for the determination of Re. This technique is well suited for samples with high dissolved solids contents, variable viscosities, high organic solvent contents or high acid concentrations, qualities typical of Re samples before anion-

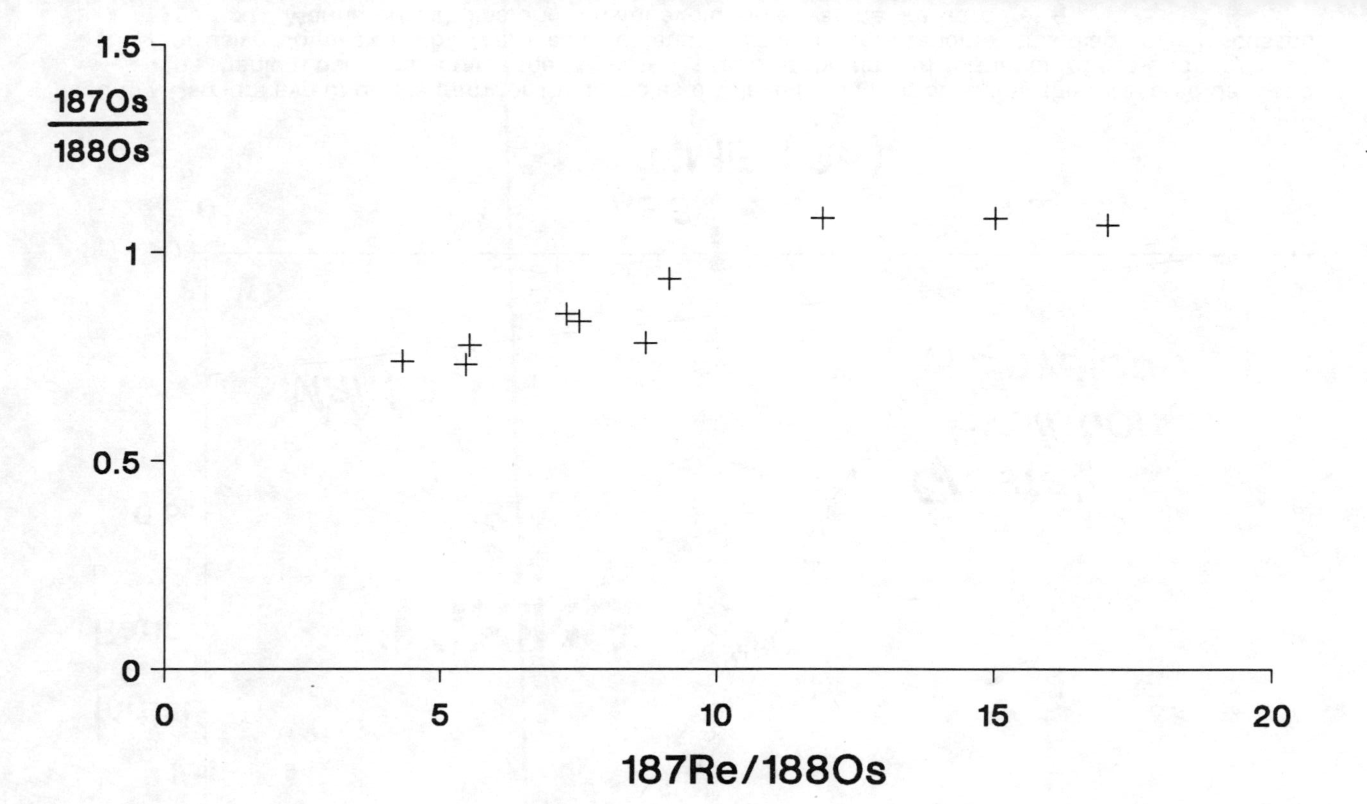

Figure 6 "Isochron-type" diagram for samples of pyrrhotite from the Creighton Mine, Sudbury, Canada[38]. If the samples were cogenetic (all formed at the same time), they would form a straight line on the diagram.

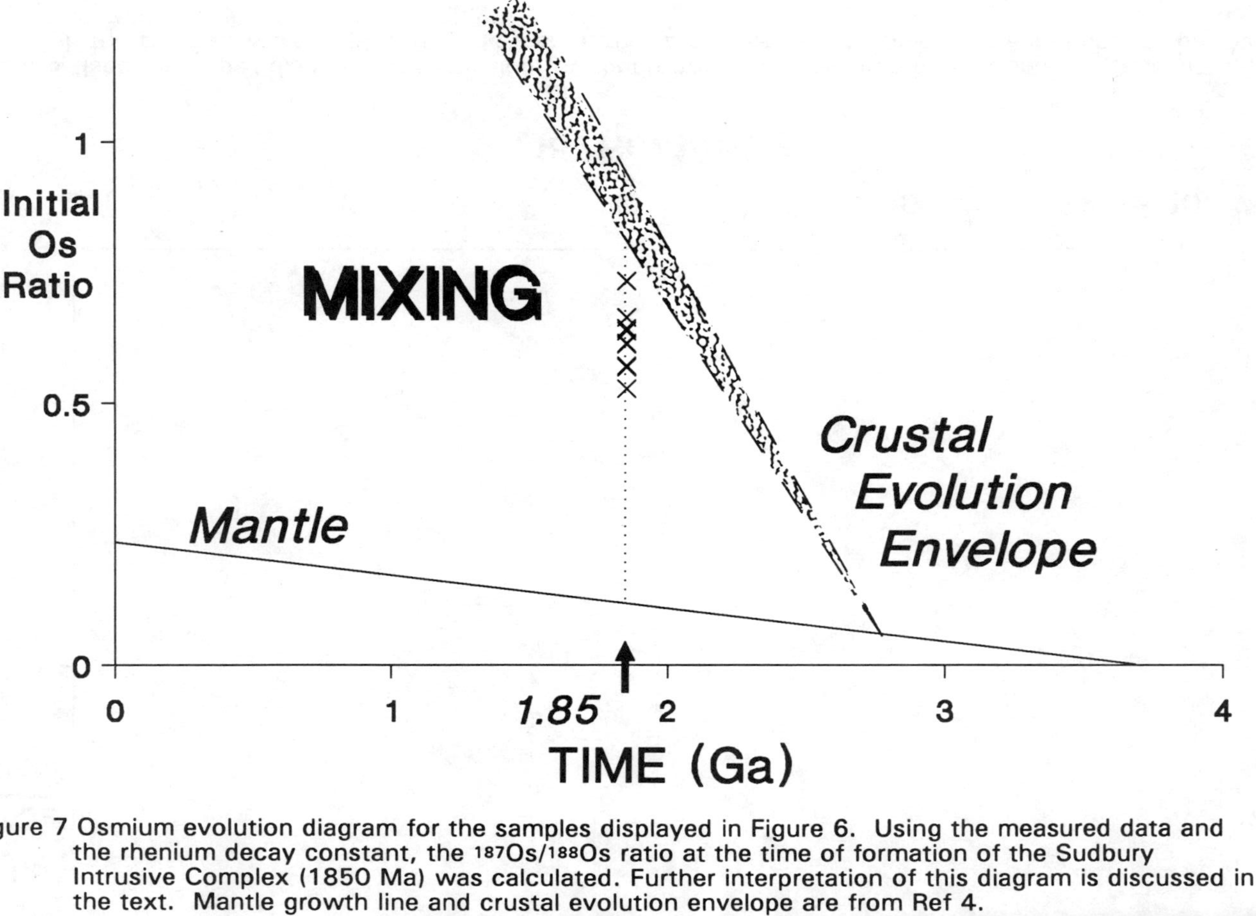

Figure 7 Osmium evolution diagram for the samples displayed in Figure 6. Using the measured data and the rhenium decay constant, the $^{187}Os/^{188}Os$ ratio at the time of formation of the Sudbury Intrusive Complex (1850 Ma) was calculated. Further interpretation of this diagram is discussed in the text. Mantle growth line and crustal evolution envelope are from Ref 4.

exchange chromatography. Small samples are used and wash-out time is low. The potential for such sample introduction techniques in ICP-MS is presently being evaluated[57]. FIA also yields a transient signal similar to that generated by the tetroxide generator, or microheater/merging apparatus. Thus data-handling techniques in this field could simplify and automate the initial steps of data handling and storage for Re and Os isotopic analyses.

The use of mixed gas (Ar-N-H combinations) and sheath gas plasmas enhances ionization across the mass range, decreases signal-to-noise ratios threefold and reduces matrix effects[58,59]. Successful application of such techniques to Re-Os determinations could potentially increase the amount of Os ionized and decrease the background, both of which would increase precision, especially on low-abundance samples.

"Multicollection" or the simultaneous collection of more than one mass enhanced the speed of TIMS analyses and improved the precision by 5%[60]. The development of multicollection capabilities for ICP-MS could similarly improve the precision of measurement of the transient Os signal.

Douglas[61] reports that the use of radio frequency biasing of samplers and skimmers controls the transmission of ions through the interface. Also the use of a curved AC-only quadrupole reduces photon noise and the use of triple quadrupole mass spectrometry (ICP-MS-MS) reduces molecular ion interferences. Space-charge effects in the ion optics can also be minimized by reducing sampler and skimmer size in samples with low dissolved solid contents. Optimial separation of the sampler and skimmers in older instruments can also improve the analyte signal by 50%, thus improving precision and detection limits[62]

8 CONCLUSIONS

The precision of Re and Os isotopic determinations by ICP-MS is probably controlled by several factors including noise in the ICP, long-term calibration and electronic stability, physical interference from sample-derived material, molecular ion effects and the current design of the instrument. These parameters impact on the counting statistics with the result that precisions better than about 0.2% on isotopic ratios are difficult to achieve. The effect of some of these parameters can be minimized by optimization studies such as those conducted by Richardson *et al.*[23,30,63] and Gregoire[21], but some are inherent in the instrument design (*e.g.* sample-skimmer separation). At present, precisions between 0.15 and 1% (2σ RSD) are routinely achieved for Re and Os isotopic ratios on natural samples containing 10 - 400 ppb Re or Os. Thus ICP-MS can produce Re and Os isotopic data that is of similar precision to that generated by other analytical techniques.

ACKNOWLEDGEMENTS

This project is funded by Natural Science and Engineering Research Council of Canada Strategic Grant STR0032713. Technical assistance by Steven B. Beneteau and James I. McAndrew was much appreciated. Discussion with Drs. J. Bazan, R. Hutton and R. Keays were very helpful. Dr. D.C. Gregoire kindly provided preprints of several papers in press. The invitation of Dr. K.E. Jarvis to address the 3rd Surrey Conference on ICP-MS and her encouragement of this review paper was much appreciated.

REFERENCES

1. M. Lindner, D.A. Leitch, G.P. Russ, J.M. Bazan and R.J. Borg, *Geochim. Cosmochim. Acta*, 1989, **53**, 1597.
2. W. Herr, W. Hoffmeister, B. Hirt, J. Geiss, and G.F. Houtermans, *Z. Naturforsch*, 1961, **16A**, 1053.
3. S. Hart, *Economic Geology*, 1990, **84**, 1651-1656.
4. C.J. Allegre and J.M. Luck, *Earth Planet. Sci Let.*, 1980, **48**, 148.
5. R.J. Walker, S.B. Shirey and O. Stecher, *Earth Planet. Sci. Let.*, 1988, **87**, 1.
6. J-M. Luck and C-J. Allegre, *Nature*, 1983, **302**, 130.
7. M. Lindner, D.A. Leitch, R.J. Borg, G.P. Russ, J.M. Bazan, D.S. Simons and A.R. Date, *Nature*, 1986, **320**, 246.
8. F.E. Lichte, S.M. Wilson, R.R. Brooks, R.D. Reeves, J. Holzbecher and D.E. Ryan, *Nature*, 1986, **322**, 816.
9. R.J. Walker and J.D. Fasset, *Anal. Chem.*, 1986, 58, 2923.
10. R.J. Walker, *Anal. Chem.*, 1988, **60**, 1231.
11. N. Shimizu, M. Semet and C-J. Allegre, *Geochim. et Cosmochim. Acta*, 1978, **42**, 1321.
12. J-M. Luck, 'PhD. Thesis: Geochimie du rhenium-osmium methode et applications', 1982, University of Paris.
13. C.E. Martin, *Earth Planet. Sci. Let.*, 1989, **93**, 336.
14. R.T.D. Teng, 'M.Sc. Thesis: Determination of osmium isotopes in meteorites and crustal samples with accelerator mass spectrometry', 1986, University of Rochester.
15. G.P. Russ, J.M. Bazan and A.R. Date, *Anal. Chem.*, 1987, **59**, 984.
16. R.S. Houk, V.A. Fassel, G.D. Flesch, H.J. Secv, A.L. Gray, and C.E. Taylor, *Anal. Chem.*, 1980, **52**, 2283.
17. A.L. Gray, Chapter 1 in Date, A.R. and Gray, A.L. eds. 'Inductively Coupled Plasma Mass Spectrometry', Blackie, London, 1989 p. 1-43.
18. R.S. Houk, *Anal. Chem.*, 1986, **58**, 97A.
19. A.R. Date, A.E. Davies and Y.Y. Cheung, *Analyst*, 1987, **112**, 1217.
20. A. Masuda, T. Hirata and H. Shimizu, *Geochem. Journ.*, 1986, **20**, 233.
21. D.C. Gregoire, *Anal. Chem*. 1990. **62** 141-146.
22. J.M. Bazan, Anal. Chem., 1987, **59**, 1066.

22a A.P. Dickin, R.H. McNutt and J.I. McAndrew, *J. Anal. Atomic. Spectrom.*, 1988, **3**, 337.

23. J.M. Richardson, A.P. Dickin, R.H. McNutt, J.I. McAndrew and S.B. Beneteau, *J. Anal. Atomic Spectrom.*, 1989, **4**, 465.
24. T. Hirata, T. Akagi, H. Shimizu, and A. Masuda, *Anal. Chem.* 1989, **61**, 2263.
25. J.C. VanLoon, *Trends Anal. Chemistry*, 1984, **3**, 272.
26. J.C. VanLoon, *Trends Anal. Chemistry*, 1985, **4**, 24.
27. R.V.D. Robert, E. Van Wyk and R. Palmer, South African National Inst. for Metallurgy, 1971, Report 1371.
28. E.A. Jones, M.M. Kruger and A. Wilson, National Inst. for Metallurgy, 1971, 1252, 11.
29. J.M. Richardson, S.B. Beneteau, A.P. Dickin and R.H. McNutt, (Abstract) Plasma Spectrometry in the Earth Sciences: Techniques, Applications and Future Trends, Kingston, U.K. 1989b. 49.
30. J.M. Richardson, S.B. Beneteau, A.P. Dickin and R.H. McNutt, in review (a), *Chem. Geol.*
31. S.B. Beneteau, 'M.Sc. Thesis: Re-Os Isotope Geochemistry of the Falconbridge East Mine, Sudbury, Ontario, Canada'. 1990 McMaster University, Hamilton, Ontario, Canada.
32. J. Davies and C.M. Du, *J. Anal. Atomic Spec*trom. 1988, **3** 433.
33. J.W. McLaren, D. Beauchemin and S.S. Berman, *J. Anal. Atomic. Spectrom.*, 1987, **2**, 277.
34. D.J. Douglas and L.A. Kerr, *J. Anal. Atomic. Spectrom.*, 1988, **3**, 749.
35. J.A. Olivares and R.S. Houk, *Anal. Chem.*, 1986, **58**, 20.
36. R.C. Hutton and A.N. Eaton, *J. Anal. Atomic Spectrom.*, 1988, **3**, 547.
37. J.M. Richardson, A.P. Dickin, J.H. Crocket and R.H. McNutt, (Abstract) 1990 Winter Plasma Conference, St. Petersburg, Fa., 1990 143.
38. J.M. Richardson, A.P. Dickin, R.H. McNutt, W.V. Peredery, S.B. Beneteau and J.H. Crocket, in review (b), *Geology*
39. A.O. Nier, Physical Reviews, 1939, **52**, 885.
40. J.W. Gramlich, T.J. Murphy, E.L. Garner and W.R. Shields, J. of Research of the National Bureau of Standards- A. Physics and chemistry, 1973, 77A, 691.
41. S.N. Naldrett and W.F. Libby, *Physical Reviews*, 1948, **73**, 487.
42. B. Hirt, W. Herr and W. Hoffmeister, 'Radioactive Dating', International Atomic Energy Agency, Vienna, 1963.
43. J.A. Payne and R. Drever,'PhD Thesis: An investigation of the beta decay of rhenium to nosmium with high temperature proportional counters', University of Glasgow, 1965.
44. R.L. Brodzinski and D.C. Conway, *Physics Rev.*, 1965, **138**, 1387.
45. D.D. Clayton, *Astrophys. Journ.*, 1964, **137**, 637.
46. D.D. Lambert, J.W. Morgan, R.J. Walker, S.B. Shirey, R.W. Carlson and M.L. Zientek, *Science*, 1989, **244**, 1169.
47. T.E. Krogh, D.W. Davis and F. Corfu, 'The Geology and Ore Deposits of the Sudbury Structure', Ontario Geol. Surv. Spec. Vol. 1, 1984.
48. R.S. Deitz, *J. Geol.*, 1964, **72**, 412.
49. W.V. Peredery and G.G. Morrison, 'The Geology and Ore Deposits of the Sudbury Structure', Ontario Geol. Surv. Spec. Vol.1, 1984.
50. T. Muir, 'The Geology and Ore Deposits of the Sudbury Structure', Ontario Geol. Surv. Spec. Vol.1, 1984.
51. H.Y. Kuo and J.H. Crocket, *Econ. Geol.*, 1979, **79**, 590.

52. A.J. Naldrett, R.H. Hewins, B.O. Dressler and B.V. Rao, 'The Geology and Ore Deposits of the Sudbury Structure' Ontario Geol. Surv. Spec. Vol.1, 1984.
53. B. Scribbens, D.R. Rae and A.J. Naldrett, *Can. Mineral.*, 1984, **22**, 67.
54. A.J. Naldrett and N.M. Evenson, 'Contamination and genesis of the Sudbury Ore', Ontario Geol. Surv. Open File Report 5643. 1987.
55. L.W. Alvares, W. Alvares, F. Asaro and H.V. Michel, *Science*, 1980, **208**, 1095.
56. W.R. Wolf and K.K. Stewart, *Anal. Chem.*, 1979, **51**, 1201.
57. J.R. Dean, L. Ebdon, H.M. Crews and R.C. Massey, *J. Anal. Atomic Spectrom.*, 1988, **3**, 349.
58. E.H. Choot and G. Horlick, *Spectrochim. Acta*, 1986, **41B**, 925.
59. D. Beauchemin, 3rd Surrey Conference on Plasma Source Mass Spectrometry. Abstract, 1989, 22.
60. K. Bell and J. Blenkinsop, *Geoscience Can.*, 1984, **11**, 50.
61. D.J. Douglas, *Can. J. Spectroscopy*, 1989, **2**, 45.
62. Doherty, W. 'M.Sc. Thesis, A practical skimmer design for inductively coupled plasma mass spectrometry: Gas dynamic and analytical considerations.' 1990 York University Toronto, Ontario, Canada
63. M.R. Palmer, K.K. Falkner, K.K. Turekian and S.E. Calvert, *Geochim. Cosmochim. Acta*, 1988, **52**, 1197.
64. B.K. Esser, K.K. Turekian, *Geochim. Cosmochim. Acta*, 1988, **52**, 1383.
65. U. Fehn, R, Teng, D. Elmore and P.W. Kubik, *Nature,* 1986, **323**, 1849.

THE FEASIBILITY OF THE USE OF ELECTROTHERMAL VAPORIZATION INDUCTIVELY COUPLED PLASMA MASS SPECTROMETRY FOR THE DETERMINATION OF FEMTOGRAMME LEVELS OF PLUTONIUM AND URANIUM

R. J. B. Hall[1], M. R. James[1], T. Wayman[1] and P. Hulmston[2]

[1]British Nuclear Fuels
Sellafield, Seascale
Cumbria CA20 1PG UK

[2]VG Elemental, Ion Path, Road Three
Winsford, Cheshire CW7 3BX UK

1 INTRODUCTION

The use of inductively coupled plasma mass spectrometry (ICP-MS) in conjunction with an electrothermal vaporization (ETV) sample introduction technique has been identified as a potential instrumental technique suitable for use in ultra-trace elemental analysis of environmental samples. Of specific importance in biological monitoring of Sellafield personnel is the determination of plutonium in urine at the femtogramme level ($1x10^{-15}$g). The technique presently employed for this analysis involves a chemical separation of the plutonium from the urine followed by solid-state nuclear track counting[1] and can achieve a total alpha limit of detection of the order of $4x10^{-4}$ Bq.L^{-1}. This is equivalent to approximately 80 femtogramme of plutonium. However, at this level the technique requires a long counting time of the order of 20 days to measure the total alpha activity.

Whilst conventional ICP-MS using pneumatic nebulisation is capable of measuring actinides, such as thorium and uranium, at the picogramme level, the technique is limited by the inefficiency of the sample introduction mechanism. The use of the ETV technique as a means to improving the sample introduction efficiency has been investigated by several workers[2-7] and would indicate that improved limits of detection at the femtogramme level are achievable. The use of ETV-ICP-MS for the determination of plutonium in urine would offer great advantages in terms of speed of analysis if adequate sensitivities could be achieved.

The purpose of this work was to assess the capability of the technique to meet a required target measurement of plutonium at a level of $2x10^{-4}$ Bq.L^{-1}. This is equivalent to approximately 38 femtogramme of total plutonium. With the isotopic composition of interest the plutonium-239 level is approximately 26 femtogramme.

2 EXPERIMENTAL

Instrumentation

The measurements were performed using a VG MicroTherm ETV furnace coupled to a VG PlasmaQuad II (VG Elemental, Winsford, Cheshire, UK) installed at Sellafield. Details of the instrument parameters are given in Table 1.

Table 1 Instrumental operating conditions

ICP-MS	
Coolant gas flow	13 $L.min^{-1}$
Auxiliary gas flow	0.5 $L.min^{-1}$
Carrier gas flow	0.6-0.8 $L.min^{-1}$
Forward Power	1350 W
Reflected Power	$<$10 W
Nebuliser	Meinhard glass concentric A type (where appropriate)
Sampling depth	10 mm
Sample cone	platinum tipped nickel, 1.0 mm
Skimmer cone	nickel, 0.75 mm
Electrothermal Vaporization	
Ar Carrier gas	0.6 $L.min^{-1}$ - 0.8 $L.min^{-1}$ varied flow
Freon gas flow	0 to 5 $mL.min^{-1}$ varied rate
Drying temp.	75 to 120°C for 2 minutes
Ashing temp.	450 to 800°C for 1 minute
Vaporization temp.	800 to 2600°C in 2 seconds
Cleaning temp.	2600°C for 15 seconds
Sample injection	10μL to 100μL

3 PREPARATION OF STANDARD SOLUTIONS

All standards and reagents were prepared from Aristar grade nitric acid, 70% v/v and deionised water "Nano-pure" of 18 $M\Omega.cm^{-1}$ quality.

Uranium

A solution of 29.9 $\mu g.mL^{-1}$ National Institute for Standards and Technology (NIST) standard reference material NIST 500 (calibrated by thermal ionisation mass spectrometry) in 3M nitric acid was used to prepare 200 ng.mL^{-1} and 2 ng.mL^{-1} stock solutions by diluting with 5% v/v nitric acid. These solutions were used to prepare 100 pg.mL^{-1}, 10 pg.mL^{-1} and 100 fg.mL^{-1} ^{235}U solutions in 1% v/v nitric acid. The uranium solutions of <1.0 pg.mL^{-1} were found to be unstable and had to be used within one hour of preparation.

Plutonium

A solution of 88.25 ng.mL^{-1} ^{244}Pu (calibrated by thermal ionisation mass spectrometry) was used to prepare a solution of 1.04 ng.mL^{-1} ^{244}Pu in 8M nitric acid. This solutions was used to prepare 10.4, 5.2, 2.6, 1.04 pg.mL^{-1}, 208 fg.mL^{-1} and 105 fg.mL^{-1} standard solutions by diluting with 1% v/v nitric acid. The plutonium solutions of <100 fg.mL^{-1} were found to remain stable for up to four hours.

4 OPTIMISATION

Instrument optimisation was carried out by nebulising a 0.2 ng.mL^{-1} uranium solution and maximising the detector response by changing the torch position, argon carrier gas flow and the potentials applied to the ion lenses. On converting to ETV mode, further optimisation was carried out by injecting 30μL of 0.1 ng.mL^{-1} mercury solution into the furnace tube and heating the furnace to 275°C and maximising the detector response by changing the potentials applied to the ion lenses. This approach gave an improved sensitivity but was difficult to use as a stable mercury signal was not always obtained.

5 EXPERIMENTAL AND RESULTS

Sensitivity and reproducibility

50 μL of a 100 fg.mL^{-1} ^{235}U standard solution was injected into the furnace, vaporized and the resulting transient signal recorded by monitoring continuously on the central channel of 235 amu (single-ion monitoring), was recorded. This was followed by a 50 μL injection of 1% nitric acid blank solution. This pattern of sample, blank, sample was repeated. The integrated peak intensities for the recorded signals and the calculated precision of the measurement are given in Table 2.

50 μL of a 1.04 pg.mL^{-1} ^{244}Pu standard solution and blank solution were alternatively vaporized and the detected signal under single ion monitoring was recorded. The integrated peak intensities for the recorded signals and the calculated precision of the measurement are given in Table 3. The calculated precision of 8% RSD for an injected plutonium concentration of 50 fg was worse than counting statistics would predict (3% RSD). This reflects the contribution to the imprecision by sources other than counting such as pipetting and variations in the volatilisation and transport of the sample material

from the furnace tube. The absolute limit of detection for both uranium and plutonium can be calculated from the data given in Tables 2 and 3 and these limits of detection are given in Table 4.

Table 2 Reproducibility of 5 fg uranium injections

Sample	Peak area (cps)
Blank 1	101
Sample 1	686
Blank 2	126
Sample 2	753
Blank 3	165
Sample 3	618
Blank 4	105
Sample 4	637
Blank 5	146
Sample 5	615
Blank 6	99
Sample 6	446
Blank 7	135
Sample 7	566

	Blank	Sample
Mean	125.3	617.3
Standard Deviation	25.1	96.2
RSD (%)	20.1	15.6

The limits of detection were calculated from the following equations:

$$LOD_1 = 3.3 \times SD_B \times C_a/I_a$$

$$LOD_2 = \text{volume injected} \times LOD_1/1000$$

$$LOD_3 = 6\times10^{23} \times LOD_2 \times 10^{-15}/\text{mass of isotope}$$

where:

SD_B = standard deviation of background counts
C_a = concentration of analyte
I_a = mean integrated peak area of analyte

Table 3 Reproducibility of 51.8 fg injections of plutonium

Sample	Peak area (cps)	
Blank 1	157	
Sample 1	1274	
Blank 2	185	
Sample 2	1077	
Blank 3	184	
Sample 3	1077	
Blank 4	177	
Sample 4	1018	
Blank 5	191	
Sample 5	1316	
Blank 6	164	
Sample 6	1112	
Blank 7	163	
Sample 7	1108	
Blank 8	170	
Sample 8	1232	
Blank 9	168	
Sample 9	1131	
Blank 10	186	
Sample 10	1082	
	Blank	Sample
Mean	174.5	1143.6
Standard Deviation	11.7	96.8
RSD (%)	6.7	8.5

Table 4 Calculated limits of detection

Isotope	Limit of detection		
	In solution ($fg.mL^{-1}$)	Absolute (fg)	Absolute no. of atoms
^{235}U	16.8	0.8	2.0×10^6
^{244}Pu	41.2	2.1	5.0×10^6

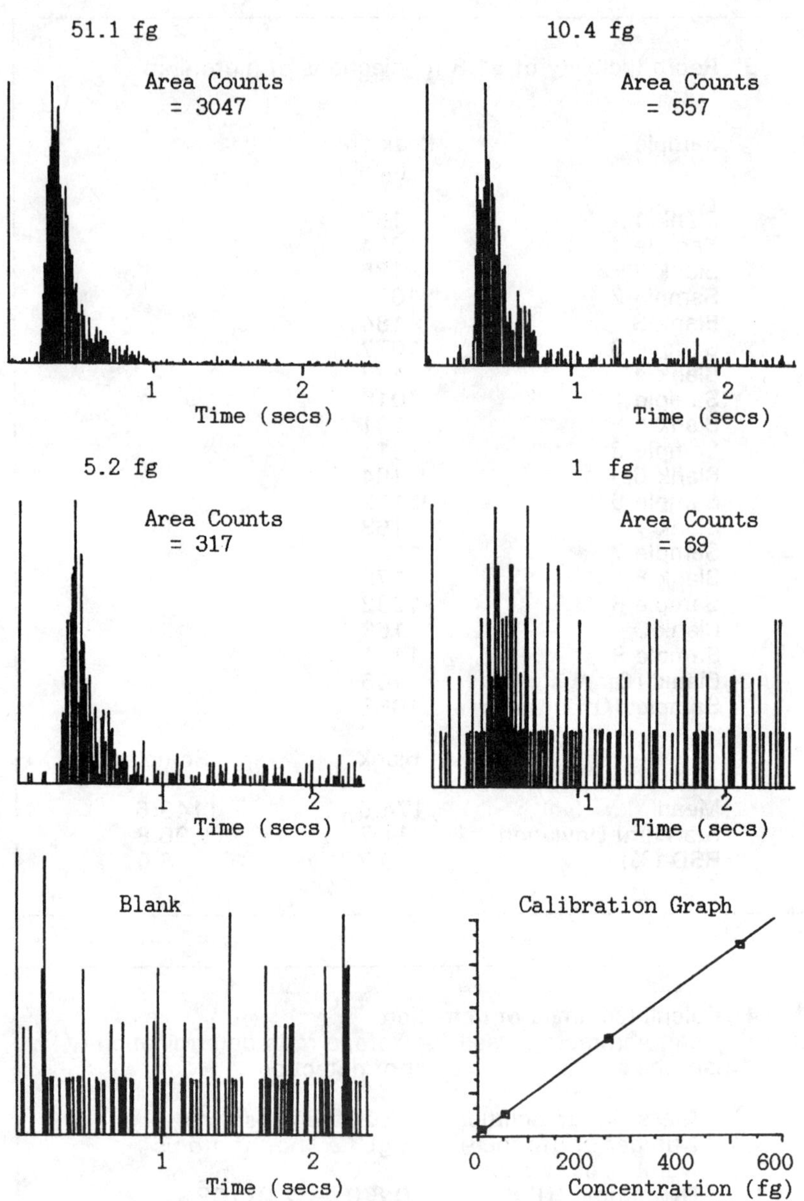

Figure 1 Scans and a typical calibration graph from the measurement of ^{244}Pu standards

Linearity

The linearity of response was measured by injecting 1, 5, 10, 50 and 100 fg amounts of plutonium into the furnace tube. Calibration data showing a best fit line and typical scans obtained from these injections are given in Figure 1.

Isotope ratio measurements

The mass spectrum from 240 to 246 amu was recorded for the plutonium-244 standard solution (50 μL injection of a 108 pg.mL^{-1} standard solution) using scanning mode. The instrument parameters for both single ion monitoring and scanning mode are given in Tables 5 and 6 respectively. A comparison of the relative integrated peak areas of the signals obtained in single ion monitoring mode with the mass peaks obtained in isotope ratio mode would suggest that isotope ratio mode is between 10 to 50 times less sensitive. The minor isotope present in the plutonium-244 solution has a concentration of 66 fg with the major isotope concentration of 5317 fg. The observed 242/244 atom/atom ratio can be seen to be 0.0138 compared to the expected value of 0.0135. The precision of this measurement was 9% RSD. Typical scans of the mass spectrum from 240 to 246 amu obtained under scanning mode are given in Figure 2.

Table 5 Parameters used for single ion monitoring detection of uranium and plutonium

Number of channels	512
Number of scan sweeps	1
Dwell time (μs)	20480
Collector type	Pulse
Mass U	235
Mass Pu	244

Table 6 Scan parameters used for isotope ratio determination of plutonium

Mass Range (amu)	242-246
Number of channels	512
Number of sweeps	500
Dwell time (s)	20
Total run time (s)	5

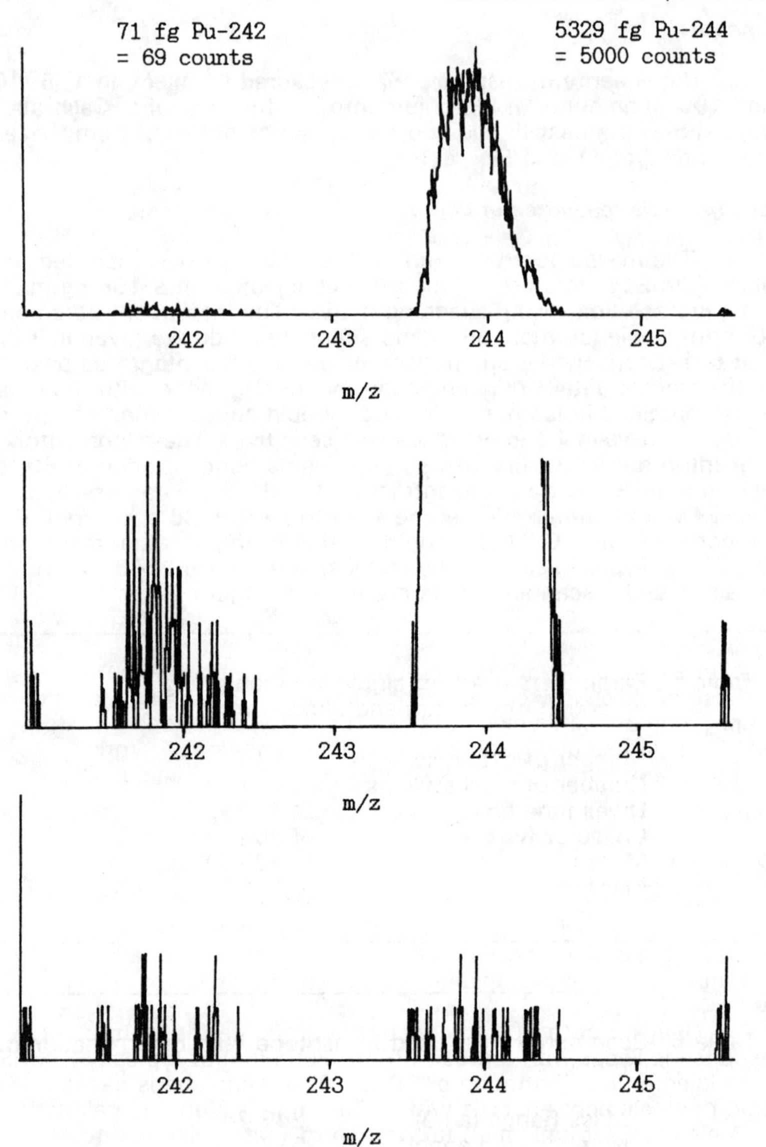

Figure 2 Mass spectra of 50μL of 108 pg.mL^{-1} ^{244}Pu solution and 50μL of 1% nitric acid blank (bottom)

6 DISCUSSION

These preliminary results have shown that the ETV-ICP-MS technique is capable of measuring both uranium and plutonium at the femtogramme level and that the target level of 26 fg of plutonium-239 can be achieved using single ion monitoring. However, for the full determination of the total plutonium alpha activity in urine it will be necessary to measure the plutonium isotopic composition. The major alpha emitting plutonium isotopes are ^{238}Pu, ^{239}Pu and ^{240}Pu with any remaining isotopes being insignificant by comparison. The measurement of plutonium-238 which is a major contributor to the alpha activity will not be possible by this technique owing to its low concentration by mass (sub femtogramme) and the effects of background levels of uranium-238 which is present in the graphite tube material and the reagents used for the analysis.

For the urine analysis it will be necessary to determine the concentration of both plutonium-239 and plutonium-240. From these measurements it will then be possible to calculate the isotopic composition of the plutonium and therefore the total alpha activity. However, as shown in Figure 2 the measurements of masses by scanning will result in a significant reduction in sensitivity compared to single ion monitoring.

The focusing of the ICP-MS instrument when operating with the ETV furnace has been found to be of critical importance when working at these extreme low concentration levels. The current procedure of evaporating a volatile element has proved to be difficult and does not lend itself to use under routine analytical conditions. Work is continuing into improving this critical area of the technique.

7 CONCLUSIONS

In single ion monitoring mode, a limit of detection of 10^6 atoms is readily achievable for both uranium and plutonium. In multi-ion monitoring a limit of detection of 10^7 atoms has been achieved for plutonium-242 and plutonium-244 measurements.

Over the 5 to 500 fg concentration range the instrument response is linear with a precision for plutonium of typically 8% RSD at an injected concentration of 50 fg. The technique is capable of rapid multi-element analysis with, after optimisation and calibration, each element determination taking approximately six minutes.

The ETV-ICP-MS technique is a powerful tool for ultra-trace areas of analysis. The sensitivity obtained in this work would indicate that the technique is suitable for the measurement of plutonium in urine. However, further work is required on improving the sensitivity when measuring ^{239}Pu and ^{240}Pu simultaneously in scanning mode.

REFERENCES

1. R. S. Grieve and T. H. Bates, *The Science of the Total Environment*, 1988, **70**, 355.
2. C. J. Park and G. E. M. Hall, *J. Anal. At. Spectrom.*, 1987, **2**, 473.
3. C. J. Park and G. E. M. Hall, *J. Anal. At. Spectrom.*, 1988, **3**, 355.
4. C. J. Park, J. C. Van Loon, P. Arrowsmith and J. B. French, *Anal. Chem.*, 1987, **59**, 2191.
5. A. R. Date and Y. Y. Cheung, *Analyst*, 1987, **112**, 1531.
6. H. Matusiewicz, *J. Anal. At. Spectrom.*, 1986, **1**, 171.
7. N. Bradshaw, 'Interim Report on ETV Development', Private Communication, January 1989.

THE DETERMINATION OF ACTINIDES IN ENVIRONMENTAL SAMPLES BY ICP-MS

J. Toole, A. S. Hursthouse, P. McDonald, K. Sampson,
M. S. Baxter, R. D. Scott and K. McKay

Scottish Universities Research and Reactor Centre (SURRC)
East Kilbride
Glasgow G75 0QU UK

1 INTRODUCTION

Authorised discharges of anthropogenic radioactivity, arising from nuclear fuel reprocessing and reactor operations, lead to the presence in the marine and terrestrial environments of low levels of the long-lived actinides ^{237}Np, ^{239}Pu and ^{240}Pu, as well as a range of shorter-lived actinides, fission and activation products[1,2]. The presence of these long-lived alpha-particle emitting actinides represents an incremental addition to the natural background of alpha radiation already present from the ^{238}U, ^{235}U and ^{232}Th natural decay series. Separation and measurement of these natural and artificial components are crucial to the understanding of their biogeochemical and environmental behaviours and hence for regulatory and radiological protection purposes.

We report here some preliminary results arising from developmental research carried out at SURRC in the use of inductively coupled plasma mass spectrometry (ICP-MS) for measuring these long-lived actinides in a range of environmental samples.

2 PRINCIPLES AND METHODS

The usual method adopted for the analysis of actinides in environmental samples is alpha-spectrometry, which involves lengthy radiochemical separation and purification steps prior to α-counting (1-30 days counting)[3]. For this technique separate and pure sources are required for the individual actinides. Samples for ICP-MS do not require such rigorous purification. A scheme, suitable for the preparation of soils and sediments for analysis by both ICP-MS and alpha-spectrometry is shown in Figure 1.

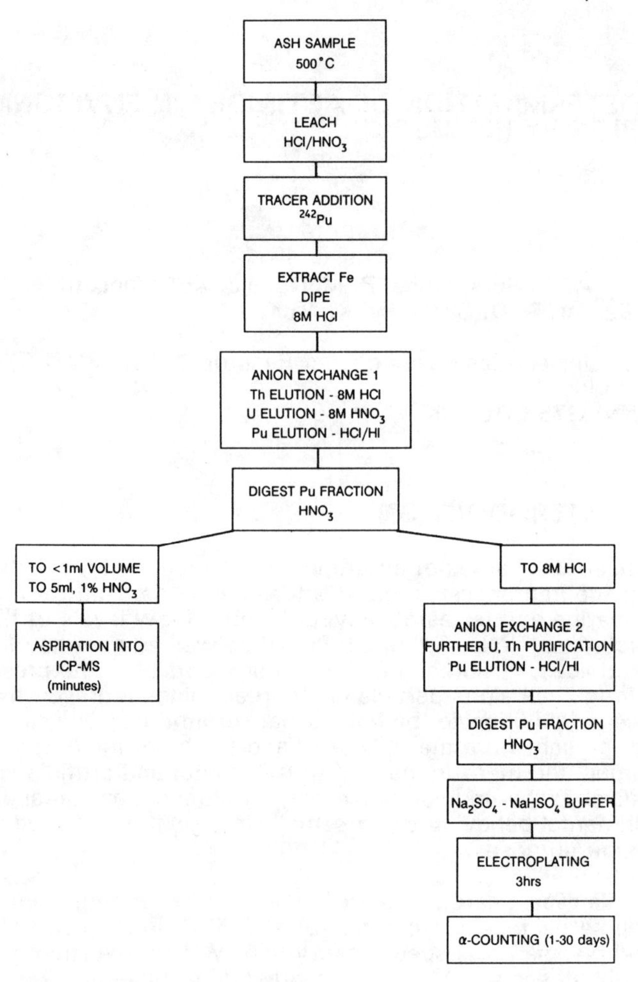

Figure 1 Flow chart showing sample preparation steps for ICP-MS and α-spectrometric analyses, illustrated for Pu in a sediment

Moreover data acquisition by ICP-MS is rapid (minutes) with the potential for simultaneous, multi-actinide assay. ICP-MS measurements are based on the detection of ions rather than decay products, and superior sensitivity is found for actinides with half-lives ≥107 years compared to α-spectrometry (Table 1). However, samples containing elevated levels of Np and Pu are suitable for analysis by ICP-MS with some form of preconcentration prior to assay. Resulting actinide mass spectra are simple, the only possible isobaric interferences in this mass range being either negligible (^{238}Pu - ^{238}U) or capable of correction by gamma-spectrometry (^{241}Am - ^{241}Pu).

Table 1 Detection limits (3σ) for actinides (μBq) by alpha spectrometry and ICP-MS. ICP-MS value based on constant background equivalent to 0.01 ng.mL^{-1}, alpha-spectrometry on a low background detector, 105s count

Actinide	Half-life (a)	α-spec.	ICP-MS
^{232}Th	1.4×10^{10}	100	0.004
^{238}U	4.5×10^{9}	100	0.12
^{235}U	7.1×10^{8}	100	0.76
^{237}Np	2.1×10^{6}	100	260
^{239}Pu	2.4×10^{4}	100	22800
^{240}Pu	6.5×10^{3}	100	84000

Table 2 Instrument operating parameters and element menus for actinide analysis

Parameter	Value
Forward, power	1375 W
reflected power	<5 W
Nebuliser	V-groove (kel-F)
Spray chamber	Scott double pass, water-cooled (13°C)
Sampling depth	10mm
Coolant flow	14 L.min^{-1} Ar
Auxiliary flow	0.4 L.min^{-1} Ar
Nebuliser flow	0.70 L.min^{-1} Ar
Sample uptake rate	0.4 mL.min^{-1} (U), 0.7 mL.min^{-1} (Pu,Np)
Mass range	113.9-116.0(Pu), 231.0-244.8(Np) 235.0-239.5(U) amu
Number of channels	512 (U), 4096 (Pu, Np)
Number of scan sweeps	600 (U), 400 (Pu, Np)
Dwell time	80 μs (U), 160 μs (Pu, Np)

Samples are introduced into the ICP in Aristar 1% v/v HNO_3 after digestion and some separative chemistry (Figure 1), freshwaters were analysed directly after acidification, and marine porewaters after dilution by a factor of 20 to reduce salt-induced signal suppression. Table 2 gives the instrument operating parameters and element menus

Table 3 Comparison of actinide results for environmental samples ($\pm 1\sigma$ mean)

Sample type	n	Analyte	Units	ICP-MS data	Reference values
Freshwater 1	(3)	^{238}U	$mBq.L^{-1}$	0.580 ±0.025	0.620 ±0.032*
Freshwater 2	(3)	^{238}U	$mBq.L^{-1}$	2.85 ±0.037	2.810 ±0.11*
Freshwater 3	(3)	^{238}U	$mBq.L^{-1}$	6.76 ±0.087	6.430 ±0.25*
Marine Porewater 1	(3)	^{238}U	$mBq.L^{-1}$	9.89 ±0.74	11.49 ±1.24+
Marine Porewater 2	(3)	^{238}U	$mBq.L^{-1}$	14.10 ±0.74	13.22 ±1.11+
Marine Porewater 3		^{238}U	$mBq.L^{-1}$	45.40 ±0.99	44.37 ±3.46+
Seaweed 1	(2)	^{238}U	$Bq.kg^{-1}$	11.50 ±0.14	10.21 ±0.32*
Seaweed 2	(2)	^{238}U	$Bq.kg^{-1}$	10.61 ±0.15	9.66 ±0.85*
Soil 1	(2)	$^{239+240}Pu$	$Bq.kg^{-1}$	0.289 ±0.006	0.311 ±0.003*
Soil 2	(2)	$^{239+240}Pu$	$Bq.kg^{-1}$	0.042 ±0.002	0.038 ±0.003*
Soil 3	(2)	$^{239+240}Pu$	$Bq.kg^{-1}$	1.039 ±0.018	1.060 ±0.040*
Mussel	(2)	$^{239+240}Pu$	$Bq.kg^{-1}$	88 ±17.00	78 ± 10*
Silt	(2)	^{237}Np	$Bq.kg^{-1}$	3.70 ±0.28 # (n = 18)	3.09 ±0.06 # (n = 4)
NRPB Sediment	(1)	^{237}Np	$Bq.kg^{-1}$	6.40 ±0.24	6.20 ±0.05**

* Alpha spectrometry of separate aliquot of material
** Result from intercomparison exercise[5]
\+ Fission track analysis
\# Weighed mean ± standard error on mean

used during these studies.

3 RESULTS

Table 3 summarises the results obtained by ICP-MS analyses of actinides in a variety of samples and compares them, where possible, to those obtained by α-spectrometry on replicate samples. Freshwaters, marine porewaters and seaweeds were analysed for U (Figure 2), soils and mussels for Pu (Figure 3) and silt and sediment for Np (Figure 4). There is excellent agreement between ICP-MS determinations and the reference values for all nuclides and for quantification using either internal standards (Pu, U) or external calibration (Np) procedures.

Attempts to determine ^{232}Th concentrations simultaneously with ^{238}U, using ^{236}U as an internal standard, have provided less satisfactory results. One problem which we have identified appears to stem from a chemical artefact, whereby Th in dilute acid (1% v/v) is partially lost from solution somewhere between the sample container and the spray chamber. This situation is not improved when selected complexing agents were added to try to retain Th in solution. However, an increase in the acid strength to 3% v/v or more did give Th/U response ratios of unity, about 3 minutes after the sample had reached the plasma (Figure 5). Accurate Th determinations using ^{236}U standard thus appear feasible at acid concentrations $\geq$3% v/v.

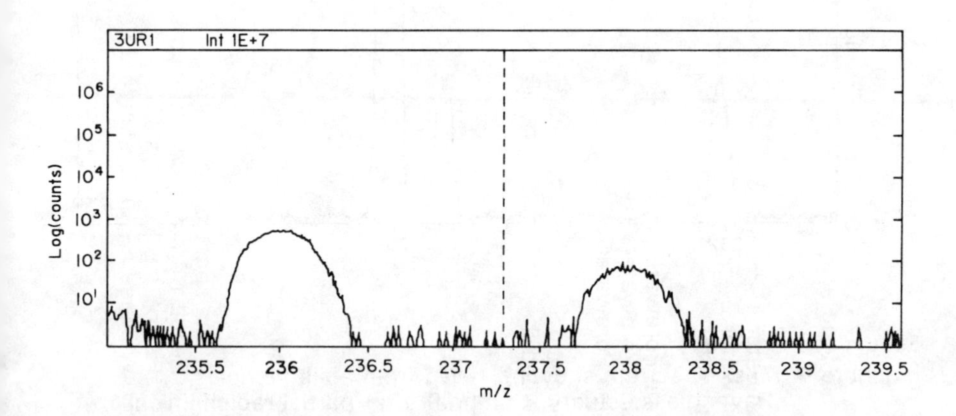

Figure 2 Mass spectrum of uranium from a sediment porewater sample. ^{236}U is internal standard, ^{238}U concentration 0.18 ng.mL^{-1} after x20 dilution.

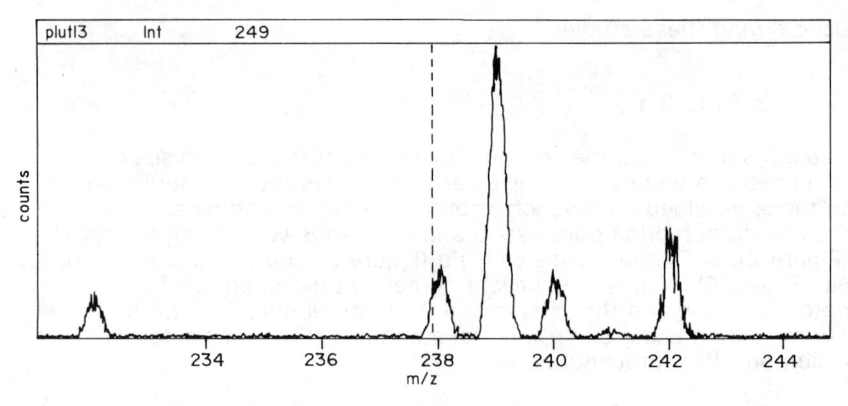

Figure 3 Mass spectrum showing ^{239}Pu and ^{240}Pu from a sample of West Cumbrian soil after partial radiochemical purification and preconcentration. ^{242}Pu has been added as an internal standard, ^{239}Pu in solution = 0.31 ng.mL^{-1}.

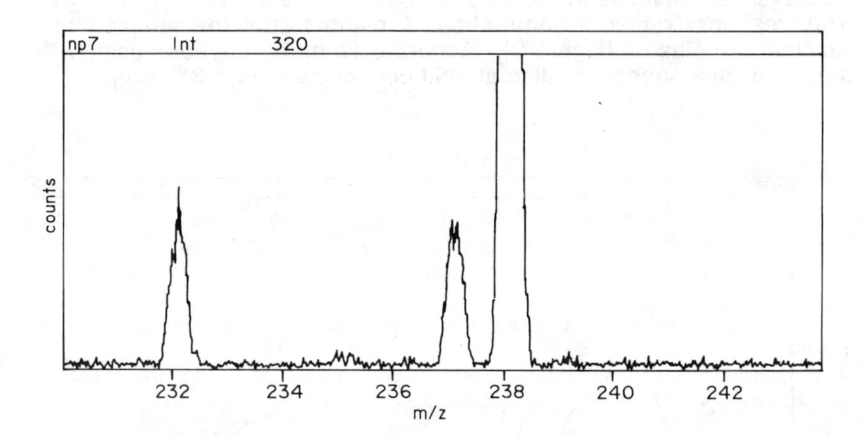

Figure 4 Mass spectrum showing ^{237}Np from a silt sample (Ravenglass estuary, Cumbria) after partial radichemical purification and preconcentration (one ion exchange plus one solvent extraction). Quantification by external calibration curve. ^{239}Np yield tracer ($t_{½}$ 2.35d) measured by τ-spectrometry prior to aspiration, ^{237}Np in solution = 0.15 ng.mL^{-1}.

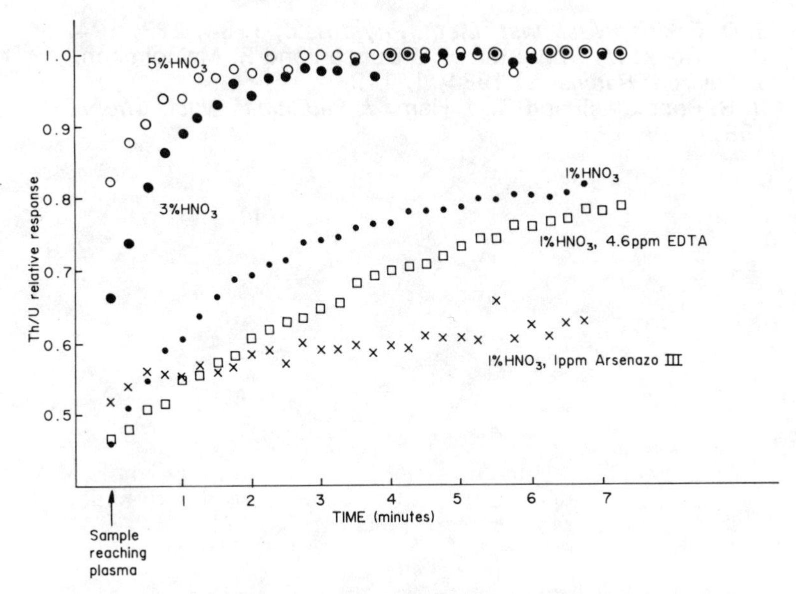

Figure 5 Relative responses of U and Th in different matrices. U and Th prepared at concentration of 10 $ng.mL^{-1}$ in all solutions

4 CONCLUSION AND FURTHER WORK

The results presented here illustrate that accurate assays of Pu, U and Np are achievable in a wide range of environmental matrices using ICP-MS. Research and development at SURRC is working towards their simultaneous determination (including Th) in environmental materials. Improvements in actinide detection limits of about two orders of magnitude by the use of electrothermal vaporisation (ETV-ICP-MS) are anticipated, and studies with this add-on facility, in collaboration with VG Elemental, have now commenced.

REFERENCES

1. F Phillips. 'Radioactive Waste Disposal by UKAEA Establishments During 1986 and Associated Environmental Monitoring Results', Safety and Reliability Directorate Report SRD R-489, 1988, 66pp.
2. BNFL, 'Annual Report on Radioactive Discharges and Monitoring of the Environment 1987' Health and Safety Directorate, Risley, 1988, 134pp.

3. J. D. Eakins, *Nucl. Inst. Meth. Phys. Res.*, 1984, **223**, 194.
4. G. T. Cook, M. S. Baxter, H. J Duncan and R. Malcolmson, *J. Environ. Radioact.*, 1984, **1**, 119.
5. D. S. Popplewell and G. J. Ham, *J. Radioanal. Nucl. Chem.*, 1987, **115**, 191.

THE APPLICATION OF INDUCTIVELY COUPLED PLASMA MASS SPECTROMETRY TO THE ANALYSIS OF IRON MATERIALS

J. A. F. Moore, M. J. McGuire and P. A. Hart

Analytical Chemistry Branch
MOD(PE)
AWE Aldermaston
Reading RG7 4PR UK

1 INTRODUCTION

The Analytical Chemistry Branch at AWE, Aldermaston provides an analytical facility available to all divisions at the establishment. As part of this service, the branch is often required to characterise materials prior to use. This characterisation includes determination of the levels at which impurity elements are present.

Several techniques are available to the branch for multielement analysis including instrumental neutron activation analysis (INAA)[1], X-ray fluorescence spectrometry (XRF)[2] and inductively coupled plasma-mass spectrometry (ICP-MS)[3]. Each of these techniques have their own limitations on the range of impurity elements which can be determined in an iron matrix and also on the concentration levels at which these impurities can be determined. This paper describes the application of ICP-MS to the analysis of high-purity iron samples.

2 EXPERIMENTAL

Apparatus

The ICP-MS used for this study was a VG Elemental PlasmaQuad. The operating conditions are given in Table 1. In order to avoid multiplier fatigue from high count rates of major species, the mass regions shown as "skipped" in Table 1 were scanned rapidly with no data being collected during this time.

Sample preparation

Samples were received as swarf. Prior to dissolution, the swarf was washed with trichloroethane to remove cutting fluid and the

Table 1 ICP-MS instrument operating conditions

Instrument		VG Elemental PlasmaQuad
Nebulisation		
	peristaltic pump	Gilson, uptake rate 1 $mL.min^{-1}$
	nebuliser	Meinhard concentric
	spray chamber	Scott double pass, water cooled to 10°C
Plasma		
	Fassel type torch	
	RF frequency	27 MHz
	Forward RF power	1.35 kW
	Reflected RF power	<10 W
Gas flows		
	coolant	13 $L.min^{-1}$
	auxiliary	0.5 $L.min^{-1}$
	nebuliser	0.5 $L.min^{-1}$
Cones		
	sampling cone	nickel; 1.0 mm orifice
	skimmer cone	nickel; 0.7 mm orifice
Vacuum pressures		
	expansion	1.9 mbar
	intermediate	$<1x10^{-4}$ mbar
	analyser	$5x10^{-6}$ mbar
Scanning conditions		
	mass range	2.0-253 amu
	skipped masses	12-22 amu 29-42 amu 53.6-54.3 amu 55.7-56.3 amu
No. of channels		2048
No. of sweeps		120
Dwell time		250 μs

surfaces leached with nitric acid (5% v/v) to remove any inorganic surface contamination. Portions of swarf (0.1 ± 0.005 g weighed to the nearest 0.0001 g) were then weighed into clean PTFE beakers and dissolved with heating, in 4 mL of 50% v/v sub-boiling distilled nitric acid[4]. Solutions were quantitatively transferred to clean polyethylene bottles and diluted with deionised water (18 $M\Omega.cm^{-1}$) to a known weight (100 ± 0.5 g weighed to the nearest 0.01 g). Indium was added to this solution at a final concentration of 100 ng.mL^{-1}, for use as an internal standard.

Standards

Iron rod (Johnson-Matthey Spec. Pure; TMI 15 μg.g^{-1}) was dissolved in nitric acid (2% v/v) to provide a 0.1% w/w matrix for standards. A mixed stock solution of the elements of interest was prepared by dilution of BDH Spectrosol standard solutions and aliquots were added to the iron matrix to provide working standards at concentrations of 10 ng.mL^{-1} and 100 ng.mL^{-1} for each element. Indium was added as an internal standard at a concentration of 100 ng.mL^{-1}.

Sample analysis

The materials studied here were:-

(a) high purity iron standard material (British Chemical Standard No. 149/3);

(b) mild steel standard material (British Chemical Standard No. 330);

(c) typical sample of iron submitted to the branch.

Portions of each of the above materials were dissolved as described above.

3 PRELIMINARY EXPERIMENTS

Matrix effects

When solutions containing matrix elements at high concentration are analysed by ICP-MS, it is common for the instrument response to differ from that observed with a simpler acid matrix[5]. This may take the form of either an enhancement or a suppression of sensitivity.

In order to assess the effect on the response of the 0.1% w/w iron matrix to be used, two standards were prepared containing the elements found in typical samples, *i.e.* Al, Mg, Ti, V, Cr, Mn, Co, Ni, Cu, Ga, As, Sr, Mo, Sn, Ba, W and Pb. The first standard was in a 2% v/v nitric acid matrix while the second contained, in addition, 0.1% w/w iron (J-M Spec. Pure; TMI 15 μg.g^{-1}).

Analysis of these two standards showed that the response in the iron matrix differed by between 10 and 30% from that in the acid alone. This difference was mostly in the form of an enhancement but some suppression was observed at the extremes of the mass range. Matrix matched standards were found to compensate for these variations and were therefore used in all further analyses.

Isobaric overlap

Another common problem encountered when analysing solutions by ICP-MS is the presence of isobaric overlaps at the isotope of interest[3]. This overlap may be due to an isotope of another element or it may result from a molecular species due to the combination of the major elements in the solution with those from the plasma gases. For example, nebulisation of a nitric acid solution in an argon plasma gives species such as ArO^+, ArN^+ and NO_2^+.

Table 2 Determination limits in the solid

Element	Determination limit* ($\mu g.g^{-1}$)
Na	200
Al	180
B	100
Mg, Br	15
Cu	10
Ti, Mn, Co, Ni, Zn, Se, Sr, Ba, Pt, Pb	1.0-10
Li, Be, V, Cr, Ga, Ge, As, Zr, Mo, Ru, Pd, Ag, Cd, Sn, Sb, Te, I, Ce, Nd, W, Hg, Th	0.1-1.0
Sc, Rb, Y, Nb, Rh, Cs, La, Pr, Sm, Eu, Gd, Tb, Dy, Ho, Er, Tm, Yb, Lu, Hf, Ta, Re, Os, Ir, Au, Tl, Bi, U	0.01-0.1

*Determination limit = 6 sd for 0.1% w/w iron solution (J-M Spec Pure; TMI 15 $\mu g.g^{-1}$)

It was observed that nebulisation of a 0.1% w/w iron matrix gave ions arising from combination of the isotopes of iron with those of oxygen and argon. The majority of these interferences occurred at masses which were not used analytically and so could be ignored.

The only problem arose in the measurement of Ge. Molecular species due to the combination of all the isotopes of iron with ^{16}O coincide with all the isotopes of Ge. Germanium is usually measured at 72 amu, but this was overlapped by the major iron oxide interference resulting from ^{56}Fe and ^{16}O, making determination of Ge below 40 $\mu g.g^{-1}$ in the solid impossible. However, use of the Ge isotope at 74 amu, which coincides with the oxide due to ^{58}Fe and ^{16}O allowed determination of Ge at below 1 $\mu g.g^{-1}$ in the solid, which was adequate for the analysis being undertaken.

Selenium was not detected in any of the materials studied, however, if high levels of this element are present the use of ^{74}Ge will be limited due to the interference from ^{74}Se (0.1% natural abundance). In such circumstances the use of an alternative isotope may give a better limit, depending on the relative levels of Se and FeO.

Table 3 Determination of impurity concentrations in two iron reference materials; a comparison of measured with reference values.

	SRM BCS 149/3 High purity iron		SRM BCS 330 Mild steel	
Element	ICP-MS	Certified[6]	ICP-MS	Certified[7]
V	<0.4	<10	280 ± 30	270
Cr	3 ± 1	<10	180 ± 20	NA
Mn	220 ± 30	190	ND	
Co	75 ± 10	70	220 ± 20	200
Ni	32 ± 3	40	110 ± 10	NA
Cu	8 ± 5	10	500 ± 50	470
Ga	<1	NA	65 ± 5	NA
As	1 ± 1	<10	40 ± 5	NA
Mo	3.3 ± 0.3	<10	70 ± 10	NA
Sn	8 ± 2	NA	50 ± 5	NA
Sb	<0.2	NA	200 ± 20	180
Pb	1 ± 1	NA	35 ± 3	30

Concentrations in $\mu g.g^{-1}$. ND = not determined
NA = no figure available

4 RESULTS AND DISCUSSION

Analysis of a solution of the iron rod (J-M Spec.Pure) allowed calculation of the determination limits for all the elements. These are given in Table 2. The analytical data for the two standard materials and a typical sample of iron submitted to the branch are shown in Tables 3 and 4 and comparison is made with reference values where these are available.

Table 4 Determination of impurity concentrations in an "in-house" iron sample

Element	ICP-MS	Others	
V	8 ± 2	11 ± 2	(b)
Cr	200 ± 30	195 ± 5	(b)
Mn	940 ± 20	900 ± 10	(b)
Co	135 ± 25	150 ± 15	(b)
Ni	320 ± 40	360 ± 10	(b)
Cu	140 ± 20	145 ± 10	(b)
Ga	44 ± 6	50 ± 2	(a)
As	23 ± 3	20 ± 2	(b)
Mo	14 ± 2	16 ± 1	(a)
Sn	9 ± 1	10 ± 1	(b)
Sb	2.3 ± 0.2	2.2 ± 0.2	(b)
Pb	1.3 ± 0.4	<5	(a)

(a) INAA (b) XRF. Concentrations in $\mu g.g^{-1}$
Errors are quoted as the 95% confidence interval

Table 5 ICP-MS precision data for 8 replicate measurements of a typical iron sample ($\mu g.g^{-1}$)

Element	Mean	SD	RSD(%)
V	8.3	0.9	11
Cr	198	16	8
Mn	940	10	1
Co	133	13	10
Ni	320	21	7
Cu	143	11	8
Ga	44	3	7
As	23	1	6
Mo	13.5	0.9	7
Sn	8.5	0.3	4
Sb	2.3	0.1	4
Pb	1.3	0.2	15

In the majority of instances, where comparison is possible, there is very good agreement. However, a very slight positive bias is observed for Mn. This may result from the incomplete resolution of the ^{55}Mn peak due to its position between the two skipped mass regions. While this bias is not significant at the relatively high levels of Mn found in the samples analysed in this study, the accurate measurement of low levels of Mn in iron could be difficult using the present scanning mode of data acquisition. Use of other data acquisition modes, such as peak jumping, may overcome this problem. The precision obtained for a typical sample is shown in Table 5. The relative standard deviation is typically better than 10% for concentrations greater than 2 $\mu g.g^{-1}$. Each of the eight portions was dissolved separately and the precision given is that for the entire analytical method.

5 CONCLUSIONS

Iron can be easily dissolved and rapidly analysed by ICP-MS. With the exception of Mn, the method shows no bias for the elements positively detected and method precision is better than 10% RSD. While interferences are present, they are largely insignificant. Analysis of B, N, O, F, Na, Al, Si, P, S, Cl, K and Ca must be carried out by alternative methods.

REFERENCES

1. R. F. Coleman and T. B. Pierce, *Analyst*, 1967, **92**, 1.
2. R. Jenkins and J. L. De Vries, "Practical X-Ray Spectrometry", 2nd Ed., MacMillan, 1970.
3. A. R. Date and A. L. Gray (Eds) "Applications of Inductively Coupled Plasma Mass Spectrometry", Blackie, 1989.
4. J. R. Moody and Ellyn S. Beary, *Talanta*, 1982, **29**, 1003.
5. S. H. Tan and G. Horlick, *J. Anal. At. Spectrom.*, 1987, **2**, 745.
6. Certificate of Analysis BCS No. 149/3 issued by Bureau of Analysed Samples Ltd.
7. Certificate of Analysis BCS No. 330 (SS No.60) issued by the Bureau of Analysed Samples Ltd.

SUBJECT INDEX